THE GILBOA FOSSILS

THE GILBOA FOSSILS

by

Linda VanAller Hernick

New York State Museum Circular 65
2003

The New York State Musuem is a program of the University of the State of New York
The State Education Department

Published 2003

Printed in the United States of America

Copies may be ordered from:

> Publications Sales
> New York State Museum
> 3140 CEC
> Albany, New York 12230
> Phone: 518-402-5344
> FAX: 518-474-2033
> Web address: http://www.nysm.nysed.gov/publications.html

Library of Congress Catalog Card Number: 2003100914

ISSN: 1052-2018
ISBN: 1-55557-172-7

Cover photo: *Aneurophyton* by Sharon Mannolini

CONTENTS

GEOLOGIC TIME SCALE

EON	ERA	PERIOD	EPOCH / Millions of Years	LIFE RECORD	Fossil (Plants / Invertebrates / Vertebrates)	Rock
Phanerozoic	Cenozoic	Quaternary	Holocene; Pleistocene 0–2; Pliocene 7	Humans, mastodons, mammoths, condors; Large carnivores		
		Tertiary	Miocene; Oligocene 26; 37–38; Eocene 53–54; Paleocene 65	Abundant grazing animals; Large running mammals; Many modern types of mammals		
	Mesozoic	Cretaceous	L; 100; E; 144	Dinosaur extinction; Earliest placental mammals; Earliest flowering plants; Climax of dinosaurs and ammonites; Great decline of brachiopods; Great development of bony fish		conglomerate
		Jurassic	L; 162; M 172; E; 201	Earliest birds; Abundant dinosaurs and ammonites		
		Triassic	L 205; M 215; E; 245	Earliest dinosaurs and mammals; Abundant cyads and conifers		
	Paleozoic	Permian	L; E; 268	Extinction of many kinds of marine animals, including trilobites; Appearence of mammal-like reptiles		
		Carboniferous	Pennsylvanian; Mississippian; 362	Earliest reptiles. Great coal-forming forests; Abundant sharks and amphibians; Large and numerous scale trees and seed ferns		conglomerate
		Devonian	L; M; E; 418	Earliest amphibians and ammonoids; Extinction of armored fish, other fish abundant		
		Silurian	L; E; 430	Earliest land plants and insects; Peak development of eurypterids; Earliest fishes in New York; Great diversity of echinoderms and ostracodes		sandstone; shale
		Ordovician	L; E; 489	Invertebrates dominant-coelenterates, mollusks and arthropods become abundant; echinoderms diversify; earliest coral reefs in early Middle Ordovician; Graptolites very abundant		limestone; dolostone
		Cambrian	L 500; M 510; E; 543	Earliest fish; Trilobites appear; Appearence of animals capable of burrowing deeply into sediments		
Precambrian Eon			ORIGIN OF EARTH 4500	Appearence of soft-bodied, many-celled animals; Single-celled algae and stomatolites; Fossils extremely rare, primitive aquatic bacteria		gneiss

Fig. 1. Geologic Time Scale.

Chapter I

INTRODUCTION

Plants furnishing the natural food for animals must have preceded animal life, yet in the earliest geologic ages, the remains of the animals are far more numerous.

> – Prof. Lucien Underwood,
> 1882

In the realm of plants there is nothing quite so striking as juxtaposing a desert setting with that of a rain forest. This contrast vividly demonstrates the diversity and complexity of modern plant life and the profound effect which the evolution of land plants has exerted at all levels, from geography to the history of life on earth . The two settings really serve to portray the earth's surface at two stages in the evolutionary radiation of plants — "before" and "after" views of the evolution, colonization, and proliferation of land plants. It was the colonization of the already ancient but barren land surface which eventually coaxed animal life out of the sea. Had plants not developed ways to tolerate the difficult transition first, the present results of life history would probably not be a subject to consider by intelligent life on earth. Indeed, the colonization of land by plants was a series of events nearly as portentous as the emergence of life itself.

PLANT EVOLUTION THROUGH TIME

Current data set the age of the earth at about 4.6 billion years. Microfossils from areas of South Africa and Western Australia provide evidence of life in the form of *prokaryotes*, simple bacteria-like organisms without cell nuclei, as early as about 3.5 billion years ago. Considering that vascular land plants, that is, plants such as trees and flowers having erect stems with water-conducting systems, first put in an appearance only some 400 million years ago, post-dating over 90% of earth's history, the obvious question becomes, "What took them so long?" This problem has been a source of speculation among researchers from many disciplines. The prevailing hypothesis is that the land environment was simply too hostile to sustain life, hence the colonization process was an extremely protracted one. Some researchers believe that the lack of a sufficiently high concentration of oxygen in the atmosphere meant that ozone was too limited in quantity to absorb the sun's lethal ultraviolet radiation.[1] It would have taken the earliest simple photosynthesizing marine organisms hundreds of millions of years to release enough oxygen to react with abundant iron and other unoxidized minerals in solution in the sea before the atmospheric concentration of oxygen was elevated to an ozone-producing, life-supporting level.

Ultraviolet radiation notwithstanding, other factors contributed to a hostile land environment. Since true soils contain an important component of decayed vegetation and active bacteria, nothing in the way of organic-rich soils would have existed on the earliest land surfaces to provide substrate and nutrients. And without plant roots to bind material produced from rocks by the process of weathering, land surfaces would have been very unstable. Deserts extended to the edges of all the seas. Dust storms were probably frequent and produced extensive dune fields. The unobstructed wind, particularly in areas with extreme variation in diurnal temperatures[2] would have greatly enhanced erosion. Rain storms would invariably lead to erosion by sheet wash and rapid cutting of desert canyons. What little soil existed in many areas would not have had the permeability to absorb water and it would have rushed down-slope and been carried away.

Instead of the deep, relatively clear rivers bordered by soils and plants that are common in the eastern United States, braided rivers like the Platte in Nebraska — shallow, shifting, interwoven streams laden with sediment in the rainy season — predominated. Most of the precipitation which fell was not retained on land but quickly returned, via the rivers, to the sea. Ponds and lakes did exist but these too were unstable environments subject to the vagaries of surrounding conditions. In a word, an important fac-

I. INTRODUCTION

tor for billions of years that explains the harshness of the early land environment is *aridity*.

What, then, was the selection pressure or evolutionary stimulus which underlay the eventual migration of life onto shore? At this point it is necessary to emphasize two aspects of the appearance of early land plants; these have been detailed by paleobotanist Dianne Edwards and her associates. First, there is the historical element — the problem of tracing colonization through geological time. A second quite separate issue is the problem of the origin and evolution of vascular land plants.[3]

In light of this distinction, it is important to note that the early land surface had actually been colonized to a limited extent well before the appearance of vascular plants about 400 million years ago. The ability to fix the sun's energy by the use of protein-based pigments and produce sugars (photosynthesis) appeared in bacteria-like forms called blue-green algae or *cyanobacteria*, the equivalent of today's pond scum. Cyanobacteria were already abundant early in the Precambrian (see Geologic Time Scale). Many of them were extremely successful, well-adapted forms which could withstand the alternating wet-dry cycles and fluctuating salinity of intertidal zones. That is, they possessed a cytoplasm, or cell contents, capable of being rehydrated and returned to normal function after drying. In this way, some were rather preadapted to life on dry land.[4] While there is no fossil evidence to provide direct proof, it is believed that a kind of scum composed of cyanobacteria and certain types of filamentous green algae was the very first covering of shore and moist near-shore surfaces.[5] As an analogue to support this suggestion, it has been observed that present-day environments, as the peaks of the Himalayas, too cold or too dry to sustain higher plants have similar microbial communities that consist primarily of cyanobacteria.[6]

Significant advances in land plants took place by 450 million years ago. Microfossils that seem to be aerially dispersed spores have been obtained from Ordovician-age rock specimens. These fossils indicate that bryophyte-like plants comparable to modern liverworts, mosses, and hornworts possibly formed a very low, mat-like ground cover in moist areas at this time. These plants probably evolved from some type of aquatic (i.e., freshwater) or terrestrial green algae.[7] As such early plant communities became more extensive and more complex, they became stable habitats for continued colonization by new groups of organisms. The subsequent emergence of microscopic detrivores (animals that eat dead plants and animals) and nitrogen-fixing plants allowed the breakdown of accumulating dead organic matter. This led to the very first nutrient-rich humus soils.

Still later, fossil evidence from Late Silurian land environments indicates the presence of centipedes and *Cooksonia*, the earliest known vascular land plant.

While hypotheses such as the environmental stresses associated with sea-level rise and fall or the existence of monsoon climates during the Early Paleozoic have been proposed to explain the evolutionary stimulus behind the initial appearance of non-vascular plant pioneers on land, there seems to be little tangible data with which to verify these opinions. One wonders whether a specific cause would even have been necessary at this earliest stage. Life is, after all, opportunistic. Given an unoccupied ecological habitat it seems practically inevitable that some organism will occupy a niche in it, a consequence of the fundamental dynamic of evolution — change.

The evolutionary pathways of vascular land plants featured changes that were a solution to the exploitation of land habitats. For example, the early land plants that invaded aquatic or swampy land habitats had no need for vascular systems within their bodies to transport water and nutrients; slow transport by diffusion provided a sufficient means of transfer. And the low-growing habit of primitive land plants confined them to an appropriately humid environment and limited their growth and propagation to wet seasons. However, plants whose bodies remained out of contact with water for extended periods, as during dry seasons, would have needed a greater volume of fluid within their tissues to overcome desiccation by wind and sun. It is believed that drying and the emergence of an erect growing habit, which may have arisen from the advantage of height in facilitating wind dispersal of spores, provided the stimulus necessary for the evolution of a vascular system.[8]

To deal with the immense morphological, physiological, and biochemical obstacles associated with transition from an aquatic to a wholly terrestrial environment, protovascular land plants produced xylem tissue which served to transport water upward. Xylem is comprised of *tracheids*, thick-walled supporting and conducting cells. *Lignin*, a wood-like substance that forms in the cell walls in the xylem and tracheids, ensured rigidity and gave some integrity to body structure.

The development of a waxy, non-cellular outer surface covering, or *cuticle*, prevented excessive leakage or drying of water from the plant and also afforded some protection against mechanical injury. Regularly spaced pores, or *stomata*, within the cuticle allowed exchange of oxygen and carbon dioxide. (This feature is a marvelous innovation that serves to open or close the plant to the atmosphere. Specialized *guard cells* around the stomata allow respiration while limiting water loss.)

Besides lignin and the cuticle, other adaptations included the development of substances to protect photo-sensitive compounds within the plant against increased exposure to light. As the evolution of terrestrial ecosystems continued, a host of compounds were developed to protect against predators that were beginning to exploit land plants as a new food source.

However, the guarantee that protovascular plants with their ever-improving structural and physiological adaptations would persist and continue to evolve lay in a key reproductive development. This was the evolution of a thick-walled spore impregnated with *sporopollenin*, a substance that made spores extremely resistant to drying and other environmental hazards. Indeed, sporopollenin is so resistant to decay that spores survive almost unchanged as microfossils in rocks up to 470 million years old. This resistant type of spore may actually have been a very early feature that appeared in some types of green algae and, as such, would have also played an extremely important role in the initial colonization of land.[9]

That we are surrounded today by a great diversity of vascular land plants is the result of hundreds of millions of years of continuous modification. This long history was marked by some major periods of diversification such as the appearance of seed plants in the Late Devonian and flowering plants during the Mesozoic Era or "Age of Dinosaurs." We are still connected to the early land pioneers by the fact that some of these species exist in our modern flora virtually unchanged from their ancestors. One example is the "shining club moss," *Lycopodium lucidulum*, that is common in present-day moist woodland areas. This plant exhibits a spiral arrangement of leaves that is nearly indistinguishable from fossil lycopods common in Devonian and Carboniferous sandstones and shales.

FOSSIL PLANTS AND THEIR PRESERVATION

As in the study of invertebrate and vertebrate animals, the evolution of plant life is studied by means of fossils. Fossils are formed in the course of the on-going *Geologic Cycle* — the process of weathering and erosion of existing rock; deposition of the resulting detritus as sediment in lakes, rivers, and seas and the entombment of dead animals and plants in these environments; compaction and cementation of the sediment into rock as a response, in part, to deep burial; and finally, eventual tectonic uplift of the new sedimentary rock ready to continue the process. Sixty-six percent of the continental surface of the earth is covered by rock which has been formed, eroded, and reformed in this way over hundreds of millions of years.[10] As long as the cycle continues there will always be areas of land exposed.

While many animals have some type of hard mineralized skeleton either outside or inside their bodies which allows them to be preserved as fossils, plants obviously do not. Actually, when one considers the many hazards encountered by a fragile plant or its parts before it can be deposited, the preservation of plant material becomes an exceedingly fortuitous event! Leaves are particularly vulnerable as they are consumed by insects and larger herbivores. If a leaf is not destroyed totally in this manner, for it to be preserved it must often be blown or carried by water to a potential site of deposition. If it does manage to arrive at a depositional site, it will again be subject to post-mortem disintegration by the action of animal detritus feeders, bacteria and fungi, and the simple exposure of dead tissue to atmospheric oxygen. It will also begin to undergo decomposition as a result of the particular chemical character of the water into which it has fallen. Should the leaf somehow survive all of these attacks and remain in a somewhat intact state as it falls to rest on a stream-, lake-, or seabed, it must be covered quickly, preferably by fine sediment for the preservation of fine detail to occur. As we have come to understand with the preservation of very finely or completely preserved animal fossils, well-preserved plant fossils often are the result of nearly instantaneous burial by such processes as local floods or volcanic ash falls. Rapid deposition of sediment at the site may produce an oxygen-depleted environment that is hostile to most microbes resulting in a higher degree of preservation.

Unarticulated or unattached plant organs — stems, leaves, roots, cones, seeds, tracheids, pollen, and spores — are by far the most common forms of plant fossils. This is attributable, in part, to the vulnerability of plants to dismemberment by wind, water, and herbivore action and to the scattering of pollen and spores to wind and water as a mode of reproduction. As a result, disassociated fossilized parts of the same plant species have often confused scientists who have sometimes given different binomial or "scientific" names to parts of the same plant.

As the type of sediment in which a plant is buried will vary and reflect the local environment, the particular type of fossilization which occurs will be a product of its setting. The most common plant fossilization process is *compression-impression*. After a plant or plant fragment is deposited and covered by a thick pile of sediment, the resulting pressure and elevated temperature often cause the cell walls of the plant to collapse and leave a black, coal-like residue. If this fossil-containing rock is split along the bedding plane, one half will contain the carbonaceous *compression* while the other will contain an imprint or *impression* without any carbonaceous material. Compressions that preserve the most detail are formed in fine sediments which give rise to siltstones, fine-grained sandstones, and shales. Many species of Eocene (i.e., early "Age of Mammals") flora are preserved as beautiful compressions in the siltstones of the Green River Formation of western Colorado, eastern Utah, and western Wyoming.

Amber, a fossilized plant resin mined in many parts of the world, preserves plant material from as early as the Carboniferous Period both as impressions and three-dimensional inclusions.

Internal structure of a plant may be preserved by another fossilization process known as *permineralization* (also called *petrifaction*). This process most often involves large plant parts such as tree trunks which are carried into a larger body of water by a river or stream where they float for an extended period, become waterlogged, and finally sink to the bottom. Another, more common process in permineralization features the rapid burial of plant debris in upland areas in the coarse sand and gravel deposited by streams in the waning stages of floods. In both processes dissolved minerals like silica, various iron oxides, or carbonates in the water are at some point precipitated from solution and remain within the cells of the plant. The hardening of this mineral matter both within and without produces a dense, rock-like infill of the original plant material. Thin sections prepared from these specimens by the use of rock saws and grinding techniques reveal the internal structure of the plant in great detail. Probably the best known examples of permineralization involve tree trunks in the Petrified Forest of Arizona (Triassic, early "Age of Dinosaurs") and the much younger (Eocene) trees in Yellowstone National Park in Wyoming and Montana.

There are two additional fossilization processes that involve a three-dimensional preservation of plants, particularly tree trunks. The first involves the production of a *mold* of the exterior of a trunk by sediment that buries the trunk. When the trunk finally disintegrates, the hollow (mold), which preserves the surface features, is left behind in the sediment. Should additional sediment be washed into this mold, or calcite or silica be precipitated in the mold by ground water, a *cast* or three-dimensional facsimile of the original trunk will be produced. Examples of this type of preservation by sand-casting are seen in the sea cliffs of Joggins, Nova Scotia, on the south side of the Bay of Fundy. Some of the Carboniferous trees are preserved at Joggins as casts in an upright, life position, a phenomenon resulting from the very rapid deposition of sand that occurs in flooding. This same process explains the preservation of the older, late Middle Devonian tree stumps at Gilboa, New York — earth's oldest fossil forests.

PALEOBOTANY

Paleobotany, the study of ancient plants, had its beginnings in Germany and Britain. As early as 1699, an illustrated catalog of fossils that included plants was printed in England by Edward Llhwyd. However, the origin of the "formed stones," as they were called, was attributed to accident or a supposedly whimsical side of nature. Only after aspects of the Geologic Cycle (discussed above) were recognized during the late 1700s as processes, and an understanding that earth history encompassed time far greater than the prevailing notion of a few thousand years, was any real progress made in the interpretation of fossils. In the succeeding two centuries, the study of plant and animal fossils throughout the world has given a fascinating backward look at our heritage. Although much of this knowledge was painstakingly accumulated due to the randomness of rock exposures and fragmentary nature of the fossil record, discoveries at a few sites with exceptionally abundant and diverse specimens and exceptional preservation have broadened our insights into certain periods of the past. Included among these windows on the past in North America are the Cambrian-age Burgess Shale of British Columbia, the Carboniferous Mazon Creek sites southwest of Chicago, the Pleistocene Rancho La Brea tar pits of Los Angeles, and the Devonian-age Gilboa Forest site in New York state.

Chapter II

THE DEVONIAN

Out of the choked Devonian waters emerged sight and sound and the music that rolls invisible through the composer's brain.

– Loren Eiseley

The *Devonian* is one of eleven time periods defined on the basis of diagnostic fossils as a fundamental interval of the *Geologic Time Scale*. It represents about 56 million years (362–418 m.y.a.) and features key events in the evolution of terrestrial life.

The *Geologic Time Scale* is really a summary of geologic history and the history of life. The scale is the result of continuous field and laboratory work over the past two centuries by geologists and paleontologists that places rocks and their fossils in the context of time. To an extent it is a contrivance because time does not come neatly packaged in intervals with definite beginnings and endings. But the fossil record within sedimentary rock around the world upon which the scale is based, while not complete in its representation, is, nevertheless, tangible proof of the passage of great expanses of time. The rocks also record such events in earth history as climate changes, sea-level fluctuations, volcanic activity, tectonic activity (i.e., regional uplifts by mountain-building), and numerous more local events which together reveal an extremely extensive earth history.

The *relative time* of deposition of particular layers of rock (or *strata*) was initially determined by early geologists in terms of older–younger based on the concept of *superposition*. This concept simply states that in a given sedimentary rock sequence, those rocks lying at the bottom were laid down first and are older than those on top. The next step was (and still is) to identify and compare the morphology or form of certain *index* or guide fossils or particular fossil assemblages within the sequence with those already known and described. In this way a more specific relative age is determined and the sequence may be correlated with similar rock and fossils in other areas. This process is known as *stratigraphy*; it includes information on the rock succession (*lithostratigraphy*) and on the fossil sequence (*biostratigraphy*).

The discovery of radioactivity near the close of the nineteenth-century afforded a new way to obtain actual age dates on rocks and determine *absolute time*. Certain elements such as uranium were found to be unstable, that is, their nuclei spontaneously disintegrate or decay at precisely determinable rates by emitting alpha particles, beta particles, or gamma radiation. This process leads to the production of new elements (daughter elements) from parent elements. Measurement of the ratio of the parent element to its daughter element for particular minerals within a rock sample can give a very accurate determination of the length of time that they have been present within the rock; this determination is equivalent to the rock's age. Techniques using the radioactive parent–stable daughter relationships of uranium-lead, potassium-argon, rubidium-strontium, and nitrogen-carbon have all contributed to the on-going refinement of age.

THE DEVONIAN PERIOD

Many of the names given to the *periods* of the geologic time scale have been derived from geographic areas in which rocks and fossils of a particular period were first recognized and described. *Devonian*, for example, is taken from *Devonshire* in southwest England. Although locally incomplete in sequence, the rocks of Devonshire provided enough evidence for nineteenth-century British geologists Adam Sedgewick and Roderick Murchison to demonstrate the uniqueness of their fossils with respect to the already described older Silurian and younger Carboniferous systems. Sedgewick and Murchison's formal presentation of the Devonshire rocks as a separate system (and hence, a separate time period) in 1839 ended nearly seven years of struggle with a problem that involved not only Sedgewick and Murchison but most of the other eminent geologists of the time. The length of the debate and the many arguments which were a part of it are peculiar to no other period of the time scale.[1] Although some later workers preferred the name *Rhenian* over Devonian, as much work on the marine faunas of the period was being done in the Ardenne–Rhenish area of Belgium, "Devonian" was accepted as a more popular term and has survived to the present day.

OTHER DIVISIONS OF THE TIME SCALE

The Geologic Time Scale is divided into three large, basic time units called *eons*: *Archean, Proterozoic, and Phanerozoic*. The Phanerozoic or "visible life" eon, which tends to have abundant (or "visible") fossils, is divided into three major units, or eras, characterized by fossils: *Paleozoic* for "ancient" life (marine life and early land plants), *Mesozoic* for "middle" life ("Age of Dinosaurs"), and *Cenozoic* for "recent" life ("Age of Mammals"). The boundaries separating these three eras mark two of the greatest extinctions in the history of life. Geological eras, in turn, are divided into geological *periods*, again distinguished on the basis of distinctive fossils. The Devonian Period lies in the middle part of the Paleozoic Era.

When referring to the rocks of a given period, or system, we use *Lower*, *Middle*, and *Upper* to distinguish relative placement within the system. But when referring to time during the period, the terms *Early*, *Middle*, and *Late* are used. As an example, tabulate corals first appeared in Early Ordovician time and are found in Lower Ordovician rocks.

THE DEVONIAN WORLD

Plate tectonic theory is the understanding that the earth's modern surface is formed of about twenty huge fragmentary plates. When these plates move away from each other they form mid-oceanic ridges where basalt is extruded from the earth's mantle to form new oceanic crust. Where the edges of the plates collide, they form mountain belts. The somewhat old-fashioned term "continental drift" reflects the origins of plate tectonic theory. Continents are the highest (and least dense) parts of the plates which rise above sea-level as land masses. It was recognized as early as 1915 by the German meteorologist Alfred Wegener that identical land plant and animal fossils occurred in South America and Africa and that these modern continents were actually part of one ancient continent. The "drift" of South America from Africa indicated by fossils and the jigsaw puzzle-like fit between the east coast of South America and the west coast of Africa was later confirmed by paleomagnetic work in the 1950s as well as by detailed mapping that showed that ancient geologic features continue from Africa into South America when the continents are brought together.

At the outset of the Devonian, the northern hemisphere was dominated by the supercontinent *Laurussia*. This giant continent included the old North American continent *Laurentia* (most of the modern U.S., Canada, northern Ireland and Scotland, Greenland, Spitsbergen, and even a sliver of northern Norway) attached to the equally ancient continent *Baltica* (Scandinavia, northern Germany and Poland, and east Russia and part of the Ukraine). The mountains that run the length of Norway represent the zone of collision of these two continents that gave rise to Laurussia. The remaining northern land mass existed as a few smaller continents.

To the south lay *Gondwana*, the largest of all the paleocontinents at this time. This supercontinent embraced Florida, South America, Africa, Arabia, Madagascar, India, Antarctica, Australia, New Zealand, southern Europe, Afghanistan, Turkey, Tibet, and Iran — more than half of the known continental crust. While Gondwana remained in a stable configuration throughout the Devonian, collisions of some of the northern continents during the period brought about several mountain-building episodes. One such collision involved a small continent called "Avalon" which collided with the eastern edge of North America. (Avalon is now "embedded" in the eastern Appalachian Mountain zone and extends from Rhode Island and eastern Massachusetts through coastal Maine and New Brunswick and northern Nova Scotia and eastern Newfoundland. Its eastern limit includes Wales, southern England, and Belgium.) This produced what is known as the *Acadian Orogeny* or mountain-building event. Sedimentary deposits from the erosion of this mountain complex formed the *Catskill Delta* of New York state.

During the Devonian, ancestral North America (Laurentia) was rotated about 90 degrees clockwise from its present orientation and was located further south because of movement of the plate. This meant that New York lay a relatively short distance below the equator. Analysis of Devonian sedimentary rock deposits in New York, particularly carbonates (limestones) from the Early and Middle Devonian, indicate that a warm, tropical climate prevailed. The northern hemisphere, comprised mostly of ocean water that absorbs solar heat, would have tended to maintain a more equable climate.[2]

The length of a day was also different at this time. Having observed that modern corals produce daily layers of carbonate, the late Cornell University paleontologist John Wells determined that Devonian rugose or "horn" corals from New York produced 400 layers of carbonate annually, hence 400 days in a Devonian year. The shorter days reflect a higher rate of the earths's axial spin.[3] Since the Devonian, tidal friction between the earth and moon has worked to decrease the rate of rotation, thus lengthening the day and leading to fewer days in the year. A further consequence of the slowing of the earth's rotation is the gradual movement of the moon away from the earth. During the Devonian, the moon was only half its current distance from the earth.[4] As a result, tides were higher along Devonian seashores.

Today, Devonian-age marine rock covers a disproportionately large continental land area. Since there is no reason to believe that this rock has been any less prone to the effects of weathering and erosion than sedimentary rock of any other age, it may be assumed that the extent of marine environments required for its formation was exceedingly great. In part, this inordinate extent of Devonian seas was a result of great mountain-building events followed by erosion and cycles of transgression of the seas onto the continents which provided catch-basins for sediment. Simply stated, Devonian sea-levels were very high. Indeed, the high sea-levels and intense mountain-building of the Devonian Period were only matched in the later part of the Ordovician, the Late Carboniferous, and the Cretaceous periods.

LIFE OF THE DEVONIAN

Devonian rocks — carbonates, sandstones, siltstones, mudstones, shales, and extensive evaporites (including salt) in western North America — represent a variety of environments or depositional settings through the period. Some of these environments were extremely extensive in their geographic range. For example, certain Devonian limestones derived from reefs are found in such wide-spread areas as northwestern Canada, eastern Australia, Spain, and Britain. A world-wide surge of evolution during the Middle and Late Devonian that involved corals, particularly rugose or "horn" corals, and other reef-building organisms such as stromatoporoids[5], was apparently stimulated by the prevalence of shallow, warm seas. Devonian rock exhibits an abundance and diversity of fossils unequaled in Phanerozoic time to that point. And when one remembers that the fossil record is very fragmentary, the abundance and diversity are even more remarkable.

With regard to marine life, not only had all major body plans of marine organisms known today evolved (this had actually occurred by the end of the Ordovician), but other important groups arose from existing groups during the Devonian Period. The *ammonoids*, a group of shelled, squid-like animals are one example. They arose during the Early Devonian from a nautiloid cephalopod group (squid relatives with external shells; e.g., the modern *Nautilus*) and became very successful. Ammonoids lasted until the end of the Cretaceous — a span of about 325 million years! Because of their incredible age range, world-wide distribution, and rapid evolution, the ammonoids have been invaluable to paleontologists as index fossils for age correlation work and as a prominent example to point to when considering some of the many problems yet to be resolved in evolutionary theory.

Other invertebrate groups which underwent significant evolutionary radiations during this time included brachiopods, echinoderms, the previously mentioned rugose corals, and some other coelenterate groups. However, the mass extinction which occurred over a three to four million period during the Late Devonian was particularly hard on the brachiopods and shallow water organisms such as tabulate corals and stromotoporoids.[6]

Because humans are biased with regard to the importance of their own species and hence the origin of vertebrates in general, the Devonian Period has traditionally been known as *The Age of Fish*. It is true that fish experienced an unprecedented period of rapid evolution during the Devonian. This included the appearance of the *Rhipidistia*, which gave rise to the first tetrapods — the amphibians and their immediate fish-like ancestors. But the *real* significance of the Devonian Period lies in the changes which occurred *overall*. The period itself was an intermediate stage, a transition from the wholly marine to the terrestrial. Of primary importance was the evolution and diversification of land plants. Late Silurian and Early Devonian plants were small and low-growing. Their lack of large penetrating root systems forced them to rely on surface water and confined their habitat to near-shore wetlands. With the development of secondary growth tissue during the Middle Devonian, plants were able to attain greater size. Large root systems, a consequence of larger size, anchored plants more deeply to the soil substrate. The combination of larger root systems and the development of the seed habit, as opposed to spores, in the Late Devonian allowed plants to extend their geographic distribution to include drier uplands. The resulting increase in biomass affected the rate, type, and volume of soil formation. Diversification of land plants also had a major impact on the earth's hydrologic cycle, e.g., in reducing runoff and increasing precipitation.[7] Ultimately, land plant evolution and diversification effected changes in global climate and, by providing detritus, afforded the first land animals a way to survive. Finally, the presence of abundant land plants set the stage for fresh-water ecosystems with diverse and myriad organisms. The Devonian, *Age of Land Colonization*, brought life to a profound new level of possibility.

Chapter III

GILBOA AND THE FIRST FOSSIL DISCOVERIES

And Time, a maniac scattering dust, ...

– Tennyson

As discussed in Chapter II, the rocks of the Catskill Delta of New York state were formed during the Devonian Period as a result of a mountain-building event called the Acadian Orogeny. The newly risen Acadian Mountains, located to the east of the present-day Catskills, were uplifted and began to erode. The resulting sediment was carried by large braided river systems westward down the slopes and across the great river flood plains toward the Appalachian Basin, a portion of the inland sea which at that time extended from Newfoundland to Alabama. This delta complex is evidenced today by its more offshore marine shales, sandstones, and minor carbonates of the Appalachian Basin and by shelf environments that are rich in invertebrate fossils. The eastern deposits of the delta — red, brown, and green shales, sandstones, and very minor coals — are terrestrial or freshwater in origin and contain fossils of plants and animals (freshwater fish and arthropods at this time) characteristic of delta and river flood plain environments.

Rocks of the Catskill Delta occur not only in New York but also in areas of Pennsylvania, Maryland, Virginia, West Virginia, and Ohio. This areal extent attests to the size of the complex.

During the younger Mesozoic Era and ensuing Cenozoic Era, eastern North America underwent successive cycles of uplift followed by erosion. After being uplifted during one of these intervals, the Catskill deposits —

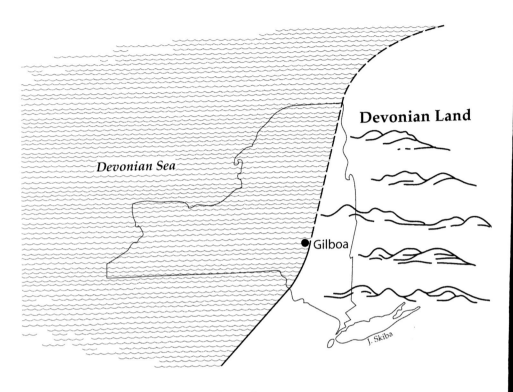

Devonian Land

Devonian Sea

● Gilboa

J. Skiba

Fig. 2. Gilboa during the Middle Devonian.

products of erosion — were themselves then subject to erosion. The Catskill deposits eventually took on the features of a "dissected plateau." What we see today as the peaks of the Catskill Mountains is regarded by some scientists as a vestige of the initial plateau surface.

The present topography of the delta complex and surrounding area is the result of the relatively recent sculpting forces of streams that were active before and after the advance and retreat of glaciers. This occurred during the Wisconsin glaciation of the Pleistocene Epoch.

GILBOA

The Schoharie Creek[1] has a source near the Plattekill Clove in the Catskill Mountains of Greene County, New York, and flows north through the entire length of Schoharie County and into Montgomery County, where it joins the Mohawk River near Tribes Hill. The course of the creek traverses about 83 miles, and its drainage basin encompasses 947 square miles with an overall relief of 3745 feet.[2] In its rocky upper reaches, the corridor of the creek is very narrow and very steep, graced by an abundance of beautiful waterfalls.

Between the points where the Manorkill and Platterkill[3] creeks enter the Schoharie from the east, a natural gorge existed where the creek was particularly suitable for mills. In 1764, this area was settled by Matthew and Jacob Dies who built the first mill.[4] The name "Gilboa" was eventually given to the place, probably in reference to Mt. Gilboa in Palestine, the site where King Saul and several of his sons were killed in battle with the Philistines. The word means "bubbling fountain"[5]; whether this was known to the early settlers and accounts for its adoption is unknown.

Construction of a variety of mills continued over the years, the largest being a cotton mill of the Gilboa Cotton Mill Company,[6] built in 1840.[7] This was a large mill for its time consisting of a four-story building with 100 looms.[8] Both the raw cotton and the resulting cotton sheeting produced by the mill were transported by steamboat on the Hudson River from Catskill.[9] Gilboa became "a trade center for the southern part of the county"[10] featuring, in addition to its mills, a tannery, a foundry, two hotels, three churches, a two-room school, a tinsmith, a blacksmith, an ice-cream parlor, a creamery, a feed store, four general stores, and a newspaper — the *Gilboa Monitor*.[11]

"Bird's Eye View" Gilboa, N. Y. "In the Catskills",

Fig. 3. Old postcard of Gilboa, New York. (from the Stone Bridge Collection of Beatrice Mattice)

III. GILBOA AND THE FIRST FOSSIL DISCOVERIES

SAMUEL LOCKWOOD — FIRST DISCOVERER OF FOSSIL STUMPS

It was to one of the three churches of Gilboa — the Reformed Church — that a recently ordained (1850) minister named Samuel Lockwood (1819–1894) came with his wife Elizabeth in December 1852.[12] Lockwood served the Reformed Church in Gilboa until August 1854 and then moved to New Jersey where he became pastor of the Reformed Church of Keyport for 15 years, resigning in 1869.[13] Having received a doctorate from New York University in 1867, he spent the rest of his career as a naturalist and educator.[14] His discovery of the tibia of the duck-billed dinosaur *Ornithotarsus immanis* in a bank along the shore of Raritan Bay in Monmouth County, New Jersey brought him to the attention of vertebrate paleontologist Edwin Drinker Cope (1840–1897), who illustrated the fossil in 1870.[15] Lockwood also discovered a backbone of a plesiosaur (a marine reptile) in Monmouth County which Cope named *Plesiosaurus lockwoodii* in his honor.[16] These fossils and those of a turtle, *Emys turgidus*, also illustrated by Cope,[17] were all found by Lockwood in Cretaceous clays and greensand which yielded many important discoveries in New Jersey in the late 1850s and 1860s.

Lockwood was a careful observer and recorder and amassed an extensive collection of fossils. Some of these are reposited today in the Rutgers University Geology Museum and the New Jersey State Museum in Trenton.

While in Gilboa, Lockwood devoted some of his time to studying the geology of the area. Because of his intense interest in natural history, he was probably familiar with the geology of New York state and was well aware of Scottish geologist Hugh Miller (1802–1856) and his discovery of fossil plants in the Devonian *Old Red Sandstone* deposits of Britain. Being knowledgeable of this work, Lockwood may actually have looked for evidence of fossil plants around Gilboa. In any case, on one of his exploratory excursions he discovered a sandstone cast of a fossil tree stump in the bed of Schoharie Creek.[18] While many fossils of small Devonian plants had already been discovered and studied in North America, particularly in Canada, Samuel Lockwood's discovery at Gilboa sometime during 1852–1854 was the first documented discovery of fossil trees in North America.

Lockwood prepared a description and drawings of his discovery and sent them to Hugh Miller in Scotland. However, Miller died in 1856 before his report to Lockwood was completed.[19]

Fig. 4. Samuel Lockwood, first discoverer of fossil stumps at Gilboa. (from *History of Monmouth County, New Jersey* – Franklin Ellis, 1885)

III. GILBOA AND THE FIRST FOSSIL DISCOVERIES

Presumably at some later time, Lockwood communicated with pale-obotanist John S. Newberry (1822–1892), who in 1866 was appointed to the Chair of Geology and Paleontology at Columbia University in New York.[20] Besides the cast of the tree stump, Lockwood found several other plant fossils. These included a small trunk that exhibited what appeared to be leaf scars and petioles. This specimen was subsequently submitted by Newberry to John W. Dawson (1820–1899) of McGill University in Montreal.[21] Dawson, one of the most famous early North American pale-ontologists, had studied land plant fossils from the Lower Devonian sand-stones of the Gaspé Bay area of easternmost Quebec and was extremely interested in what was being found in the Devonian strata of New York. Lockwood's smaller specimen was named *Caulopteris lockwoodi* and illus-trated by Dawson in the *Quarterly Journal of the Geological Society of London* in 1871.

FOSSILS AND THE FLOOD OF 1869

Schoharie Creek has had a long history of flooding. Some 17 cata-strophic floods are recorded since 1784.[22] These floods have regularly destroyed buildings, bridges, roads, and crops. The most recent floods of 1987 and 1996 caused more than $65 million in damages in five counties and claimed the lives of a dozen people and many farm animals.

Snow-melt from the Catskill Mountains, along with a tendency for cer-tain areas such as Tannersville in the creek's upper reaches to receive excessive amounts of rainfall, cause many feeder streams of the creek to reach capacity quickly. In this way, Schoharie Creek can become a devas-tating force practically over night, despite the two large reservoirs that have been constructed along its upper course.

In early October of 1869, a flood of this kind destroyed the mills along the creek at Gilboa (including the big cotton mill), wiped out bridges and roadbeds, and damaged many other businesses and homes in the vil-lage.[23] While quarrying rock along the creek just south of the bridge in Gilboa for use in the ensuing bridge and road repairs, workmen discov-ered some *in situ* (in life position) fossil tree stumps. (The stumps were actually exposed as a result of blasting, not by flood water as has some-times been reported.[24]) A letter received by James Hall (1811–1898), State Paleontologist and first director of the State Museum of Natural History, from "D. Mackey" of Gilboa and dated January 3, 1870, describes the discovery and removal of two stumps after the blasting. The letter states that they are "preserved for Mr. Hall to place them in the State Geological Hall at Albany."[25]

Presumably "D. Mackey" was Daniel Mackey, a resident of Gilboa who was baptized and became a member of Samuel Lockwood's former parish in 1855, shortly after Lockwood left.[26] Mackey mentioned Lockwood's earlier discovery of a fossil stump in his letter to Hall. Conceivably, Mackey and Lockwood were well acquainted. Mackey may have been more than a little interested in Lockwood's discovery, or maybe he was simply acting at the urging of others. In any case, Mackey knew the paleontological value of what was found at Gilboa through Lockwood and cared enough to secure the stumps (with the help of "Dr. Layman, Mr.Baldwin and Mr. Stryker"[27] of Gilboa) and informed James Hall. Mr. Baldwin (W.L. Baldwin) donated additional plant fossils to the Museum on his own in 1871.[28]

That Samuel Lockwood failed to notify Hall when he made his initial discovery and removed his collection of Gilboa fossils to his residence in Keyport, New Jersey, may seem rather surprising. One can only assume that this was simply because Lockwood knew that no one in Albany was any more knowledgeable about Devonian plant fossils at that time than he was. As mentioned above, the pioneering work on the Devonian flora was just underway, not in the United States, but in Canada, Britain, and central Europe.

To say that James Hall was excited by the Gilboa discovery is an understatement. Immediately after receiving Mackey's letter, he dispatched a member of the museum staff, Joseph Lintner, a resident of Schoharie (later appointed State Entomologist), to bring back several trunk fragments. Hall visited the site shortly after this to evaluate the position of the stumps within the local rock sequence and brought three of them back to Albany. Hall noted the gray sandstone in which the stumps were embedded and the shale which surrounded their bases. He concluded that "the condition of the strata, which are gently dipping to the southeast, leave no doubt that these trees had grown in the position and in the places where they were found."[29] He then sent an assistant to the site to secure any additional specimens that might turn up as the work progressed. A total of three stumps were brought to Albany along with many fragments and other plant specimens.[30] Hall had the stumps placed in "a conspicuous position at the east end of the first floor of the museum." [31]

III. GILBOA AND THE FIRST FOSSIL DISCOVERIES

The discovery was mentioned at the January 1870 meeting of the Albany Institute and was described in the January 30, 1870 edition of the Albany *Argus* newspaper. In 1872, Hall brought these fossils to the attention of the *British Association for the Advancement of Science* which met at Brighton.[32] The Brighton meeting gave him an opportunity to make his first trip to Europe.

The Gilboa collection was sent by Hall to Dawson in Montreal, who thought the stump specimens represented two different species of tree fern — *Psaronius textilis* and *Psaronius erianus*. This conclusion was based on what Dawson interpreted as the aerial roots of the stumps. These specimens were part of Dawson's illustration of the two species in his 1871report in the *Quarterly Journal of the Geological Society of London*.[33]

It should be noted that, although this 1870 discovery and that of Samuel Lockwood between 1852 and 1854 were the first discoveries and report of fossil tree stumps, the existence of fossil plant material in the area of Gilboa had been known for some time.

The New York State Geological Survey was organized in 1836 through a provision of the state legislature to survey the mineral resources of the state. New York had been divided into four geological Districts with a separate geologist in charge of each District. The First District included the Catskills and was under the supervision of William W. Mather, a retired army lieutenant who taught geology and chemistry at West Point.[34] In his report of 1843, Mather included figures of plant fossils — fragments of stems and leaves — from rock sections which he described at Manorkill Creek and Schoharie Creek near Prattsville.[35] This was the very first formal report of plant fossils in the vicinity of Gilboa (Fig. 5).

Plant fossils continued to be found around Gilboa after 1870. At that time controversy surrounded the question of whether the Upper Devonian deposits of New York state were equivalent to the Old Red Sandstone of Britain. James Hall, as early as 1844, had largely convinced himself that indeed they were, based on the apparent presence of the Old Red Sandstone in Blossburg, Pennsylvania.[36] But, other geologists disagreed, and as late as 1870 the question had not been resolved. The issue bothered Hall to such a degree that in 1871 he hired Andrew and Clark Sherwood of Mansfield, Pennsylvania. These geologists were familiar with the Old Red Sandstone in their area, and they were assigned to produce maps of characteristic outcrops in Pennsylvania, the Catskills, and adjacent sections to the west. The project continued for four summer field

CATSKILL DIVISION.

Fig. 3.

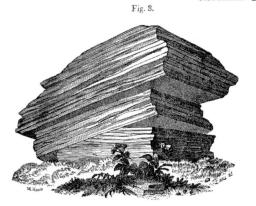

Fig. 9.

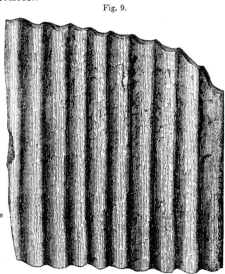

This illustrates the laminæ of deposition so strongly marked in the sandstones of the Catskill division. (Vide also Pl. 6, figs. 3 and 4.)

Fig. 10.

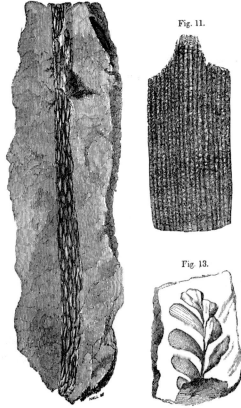

Fig. 11.

Sigillaria simplicitas (Vanuxem). This fossil differs from that of the Coal formation, in having no markings on the elevated parts.

Fig. 12.

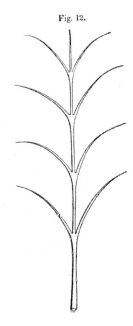

Fig. 13.

Fig. 5. Plant fossils from the Catskills — illustration from William W. Mather's *Geology of New York, Part 1*, 1843.

III. GILBOA AND THE FIRST FOSSIL DISCOVERIES

seasons under Hall's personal funding. By 1875, the evidence in support of his belief was conclusive.[37]

While mapping in the area of Gilboa at Manorkill Falls the Sherwoods found "stumps, leaves, stems" in a section of gray sandstone below the base of the falls.[38] Some of these fossils were brought back to Albany and were added to the Museum's collection.

During the summers of 1895 and 1897, the State Geological Survey undertook a comprehensive survey of the stratigraphy of the rocks included in the Hamilton and Chemung Groups in eight counties in eastern New York. The field region included Schoharie County and Gilboa. The work was done by Charles Smith Prosser (1860–1916), professor of geology at Union College, and a group of his students. Aware of the plant fossils around Gilboa and of Hall's interest in them, particular attention was given to the region in this regard. As a result, several small stumps and fragments of stumps were found lying along the road near Manorkill Falls.[39] Evidently, they had been loosened from the sandstone (Middle Devonian – Moscow Formation) in which they stood when excavations were made for the bridge foundation over the falls. No one involved in that enterprise apparently either knew what they were or felt inclined to report them, much less retrieve them. The stumps were enthusiastically collected by Prosser who also appreciated the fact that this locality was higher in elevation than the original Gilboa site and lay higher in the rock sequence.

In a bluish shale in the gorge at Gilboa bridge, Prosser found another type of plant fossil which he sent to David Pearce Penhallow[40] who had been appointed by Dawson to the Chair of Botany at McGill University in 1883.[41] Prosser also visited the original Gilboa site of 1870 from which the first stumps had been removed. He remarked that, "the impression of the side of one of these trunks still shows very clearly in the sandstone at this locality."[42]

GILBOA DAM AND RESERVOIR

A research hiatus of some twenty years followed Prosser's work. The twentieth century had arrived and with it the problems brought about by industrialization and expanding urbanization, particularly in New York City. The municipalities surrounding the harbor of Old New York City were consolidated into Greater New York City in 1898. This consolidation meant the creation of a giant city with a population of about 3.25 million people.[43] The whole New York City area had been experiencing deficiencies in the existing water supply which utilized water from the Croton River. The Croton reservoirs and aqueducts had been in use since 1842.[44] By 1904, the population of the city had increased to 4 million and was continuing to grow at a rate of about 115,000 people per year.[45] The water supplies of all the boroughs were simply inadequate in terms of quality and quantity at this level of population growth.

A committee was appointed by the city to consider the water problem in 1897. The ultimate solution was determined to involve the tapping of another, greater watershed, and the Catskills, lying just to the north, were recognized as an area that could provide this resource. By damming and impounding the waters of the Esopus and Schoharie creeks and diverting them south by means of an elaborate conduit system, an adequate and reliable supply of good water with a capacity of 148 billion gallons[46] would be assured for New York City.

In 1905, an act of the New York State Legislature known as *Chapter 724 of the Laws of 1905* was passed. The specific title of *Chapter 724* was "An Act to provide for an additional supply of pure and wholesome water for the City of New York and for the acquisition of lands or interest therein and for the construction of the necessary reservoirs, dams, aqueducts, filters and other appurtenances for that purpose;..."[47]

By June 1914, the Ashokan dam was essentially completed and the massive reservoir was beginning to store water from the Esopus Creek watershed. In October of that year, plans to utilize the Schoharie Creek watershed were approved by the State Water Supply Commission.[48] The gorge at Gilboa was found to be an excellent site for a reservoir and final approval of a plan was given by the State Conservation Commission in June 1916.[49] Construction of the Gilboa dam and reservoir commenced in 1917 with the first public notice of land parcels to be acquired by the Board of Water Supply. This notice was given in a meeting held on October 20, 1917, in the court house in Troy, New York.

After acquisition by the Board of Water Supply, the buildings of the entire village of Gilboa were either razed or burned. The three local cemeteries were moved with the reinterment of 935 bodies.[50]

Most of the residents of Gilboa were, understandably, frustrated and indignant over the loss of their property and their security. In hearings held in Kingston, New York, claimants fought both for what they believed

III. GILBOA AND THE FIRST FOSSIL DISCOVERIES

to be the true value of their property and for some sort of compensation for loss of employment and inconveniences suffered by the damming of the creek. Those people living downstream from the dam were also affected after 1926 when the dam was sealed. Problems stemmed from the diminished flow of the creek. For example, ice could no longer be harvested.[51] In addition, the creek now was too shallow and dirty to bathe in,[52] and fish populations decreased.[53] Farmers in the Schoharie valley complained that less water was available during the summer months for crops and livestock.[54] A claimant in Esperance, about 30 miles downstream, testified that his revenue from renting boats and campsites to fisherman during the spring and summer months was severely diminished by a decrease in the flow of the creek.[55]

There are about 50 volumes of claimant testimony that were found in the basement of Schoharie County's former municipal building.[56] These volumes are now reposited in the archives of the County Clerk's office in Schoharie. Some of these volumes afford a very personal and extremely detailed perspective on the importance of a creek in the lives of rural people in the early part of the twentieth century.

WINIFRED GOLDRING'S CONTRIBUTIONS

Winifred Goldring, paleontologist and paleobotanist, was born February 1, 1888, in Kenwood, New York, the fourth child in a family of eight daughters and one son. A gifted student, she graduated from Albany's Milne School as valedictorian in 1905 and received both a B.A. with honors in 1909 and a M.A. in 1912 from Wellesley College. While originally intending to major in classical languages, her interests turned to botany and zoology. She also took courses in geology and geography. Additional graduate work in geography, paleontology, and paleobotany was done at Harvard University, Columbia University, and Johns Hopkins University respectively.

After holding teaching positions at Wellesley and the Teachers' School of Science in Boston, she was hired by the New York State Museum as a Special Temporary Expert in Paleontology from 1914 to 1915. She then became an Assistant in Paleontology but resigned briefly in 1918.[57] This resignation presumably took place because of insufficient salary and pressures from both exhibit work and her 670-page monograph on the *Devonian Crinoids of New York* that had been undertaken in 1916 at the

Fig. 6. The razing of Gilboa village — dam construction in the foreground.
(courtesy of the Schoharie County Historical Society)

III. GILBOA AND THE FIRST FOSSIL DISCOVERIES

Fig. 7. Main Street, Gilboa — dam construction in the background. (courtesy of the Schoharie County Historical Society)

Fig. 8. Dam construction at Gilboa. (from the Stone Bridge Collection of Beatrice Mattice)

Fig. 9. Gilboa dam, 1928.

Fig. 10. Winifred Goldring — Wellesley College graduate, 1909. (courtesy of Mrs. Marion Goldring)

Fig. 11. Winifred Goldring. (courtesy of Mrs. Marion Goldring)

request of State Paleontologist and Museum Director John Mason Clarke (1857–1925). However, she returned to the museum and remained an Assistant in Paleontology until 1925. From 1925 to 1928, she held the position of Paleobotanist, a consequence of the museum's burgeoning collection of plant fossils. It was in this position that her name became well-known. The following years saw her as Associate Paleontologist, Assistant State Paleontologist, and finally, in 1939, after receiving an honorary doctorate from Russell Sage College in 1937, as State Paleontologist. Winifred Goldring was the first woman in the world to hold this kind of position, and she retained it with much distinction until her retirement in 1954. She was well-known both nationally and internationally for her work. As John M. Clarke once remarked, she was a "particularly hardy seeker after truth."[58]

In 1919, the New York State Geological Survey began dispatching collectors to prospective areas of the state in a serious search for fossil plant material. This project was undertaken in response to an effort on the national level to establish paleobotany as an important part of geology. Some of this work was done in central and western New York. Vincent Ayers collected in quarries and explored outcrops near Monroe, New York.[59] Jacob Van Deloo and his assistants collected near Olean, Wellsville, and Almond. In the eastern part of the state Winifred Goldring and her assistant, Joseph Bylancik, collected in areas near Walton, Cadosia, Hancock, Shinhopple, Downsville, Sydney, and Oxford. Goldring and D. Dana Luther also collected in western New York near Naples and Honeoye.[60] Much excellent material was obtained from these sites.

However, Gilboa became the focus of particular attention because of the earlier discoveries there. In addition, the onset of dam and reservoir construction imposed a time constraint; soon the excavations would be under water.

During the 1920 field season, Herbert S. Woodward, Rudolf Ruedemann, and Chris Hartnagel of the Geological Survey went to Gilboa to relocate or find new evidence of the fossil forest that yielded the first fossil stumps. Their efforts were rewarded by the discovery of new material in situ. These specimens also rested on the typical bed of dark shale, but lay at a higher level (1120 feet) along the road near the lower falls of Manorkill Creek[61] (Fig. 12). This was the area where Charles Prosser had found the small, loose stumps in 1897. On the same trip, Ruedemann discovered what were thought to be seeds associated with the stumps:

III. Gilboa and the First Fossil Discoveries

"Shortly before, I had seen figures of Carboniferous seed ferns described by David White, and had learned from Miss Goldring that Professor Johnson, of Dublin, had predicted Devonian seeds. So I thought I might look for them in the Devonian plant beds at Gilboa, but no promising black shale could be found. One night, as we were trying our luck at fishing in the Schoharie Creek, both my companions went out on a big bowlder [sic] in the river while I remained on shore to enjoy the beautiful scenery of the Manorkill Falls just behind us. While I was sitting there, I noticed a large slab of black shale sticking out of the river sand. It was covered with beautiful clusters of fern seeds. I seized it, and yelled to the fishermen. We carried the precious slab home. When we arrived, my companions discovered that they had left their pipes and tobacco on the bowlder and blamed me for having made too much 'fuss' over my find. But it was worth it, for these are the oldest seeds known at present. There was no doubt that the bed from which the slab came was near by, and we found it the next morning not a hundred feet away between the Manorkill and Schoharie creeks. From it we secured a fine collection of seeds. Miss Goldring went out later and obtained the sporebearing organs, the foliage and rootlets."[62]

Quarrying for the dam, which was being built directly on the site of the 1870 discovery at an elevation of 1020 feet, was progressing, and seven additional fossil tree stumps were found.[63] Even more exciting discoveries were made in a quarry along Schoharie Creek just north of the original locality (Figs. 16, 17). Construction workers uncovered a bonanza of fossil stumps in Riverside Quarry (Middle Devonian – Moscow Formation) at an elevation of 960 feet. Eighteen were removed from a 50-square-foot area. The largest of all the stumps discovered at Gilboa, with a basal circumference of eleven feet, was found here.[64]

The addition of these new specimens to the New York State Museum's collection brought its total to 40.[65] They represented three successive forests that had grown along the swampy, tropical coast of a Devonian sea which periodically rose to flood and eventually bury each successive forest.

Fig. 12. Fossil tree stump *in situ* at the highest tree elevation — 1120 feet — along the road above the lower falls of Manorkill Creek. Photo by Chris A. Hartnagel, 1920.

Fig. 13. Lower falls of Manorkill Creek where "seeds" were found by Rudolf Ruedemann in 1920. Photo by Chris A. Hartnagel, 1920.

Fig. 14. Bridge over the falls of Manorkill Creek. (from the Stone Bridge Collection of Beatrice Mattice)

III. GILBOA AND THE FIRST FOSSIL DISCOVERIES

Fig. 15. Stone supports of the old bridge (Fig. 14) over the lower falls of Manorkill Creek, revealed during a period of low water in the reservoir, 1991. (photo by the author)

Many more badly weathered stumps were discarded. Others were destroyed by blasting, and several excellent additional stumps were sent to museums by the ever-accommodating Hugh Nawn Contracting Company, builders of the dam.

Engineers J.A. Guttridge, Thaddeus Merriman, George G. Honness, and Sidney K. Clapp of the New York City Board of Water Supply[66] also made every effort to keep in touch with the Museum during the dam and reservoir construction. They apprised Winifred Goldring and John Clarke of new stump discoveries as they occurred and retrieved other plant fossil material from the many spoil heaps around the site. Without their vigilance, many of the fossils would most certainly have been lost.

In December 1920, Clarke sought permission and funding from the State to send Winifred Goldring to Johns Hopkins University for a semester of study with the preeminent paleobotanist Edward W. Berry.[67] Both petitions were granted, and from January to May of 1921, Goldring added a knowledge of fossil plants to her already extensive botany background. On her return to Albany, she began the awesome task of describing the wealth of Gilboa material, "the most impressive display of the early land flora of the known world."[68] Much of this material had been put on display in the central hall of the State Museum.

As mentioned above, William Dawson thought that the stumps had an affinity with the genus *Psaronius*, a tree fern common during the Carboniferous and Permian periods. His conclusion had been drawn from an examination limited to the stumps. But, much additional plant material had been found at Gilboa since Dawson's death in 1899. There were what appeared to be isolated seeds and micro-sporangia, and fronds or branch systems up to 1.8 meters (six feet) long. Because these separate fossils often occurred in close association in the black shale at the base of the stumps Goldring naturally assumed that they all derived from the same plant. She proceeded to reconstruct the plant based on this supposition and on her belief that the aerial roots of the stumps described by Dawson were actually a type of sclerenchymatous (thick, tough) outer cortex or bark quite different from *Psaronius*.[69] Goldring concluded that the Gilboa fossils were arborescent *seed ferns (pteridosperms)* that reached an average height of 6–12 meters (20–40 feet), had a bulbous base like that of a cypress, and long tripinnate (having three divisions of the branches with leaves pinnate or arranged in a feather-like pattern) branch systems. The seeds apparently occurred in pairs near the tips of the branch systems.

III. GILBOA AND THE FIRST FOSSIL DISCOVERIES

Dawson's *Psaronius erianus* and *Psaronius textilis* were referred by Goldring to a new genus *Eospermatopteris* (i.e., name meaning "dawn of the seed fern"). She still recognized Dawson's two species, but these were now named *Eospermatopteris erianus* and *Eospermatopteris textilis* respectively (Figs. 20, 21).

During a behind-the-scenes visit to the American Museum of Natural History in 1919, Goldring observed the work of the museum artists and "got a few suggestions." She said, "I like their line drawings better than wash drawings and I am sure I can do it with a little practice."[70] And practice she did. Her beautiful reconstruction drawing of *Eospermatopteris erianus*, the "Gilboa tree," for her paper, *The Upper Devonian forest of seed ferns in Eastern New York* published in 1924, has been reproduced in hundreds of books and articles ever since (Fig. 23).

Goldring initially thought that the plants she was dealing with had a relationship to *Lyginopteris*, a genus of vine-like Carboniferous seed ferns described from fossils in England. These plants had fern-like foliage with seeds on the fronds, and the compression surface showed an anastomosing outer cortex[71] like one of the species of Gilboa plants. On the basis of these morphological similarities, particularly the outer cortex, she considered proposing *Protolyginopteris* as a new genus name for *Eospermatopteris*. Her mentor E.W. Berry commented, "I am delighted that the Gilboa pteridosperm is coming along so nicely. I can see nothing the matter with the name *Protolyginopteris* if you are quite sure of the relationship."[72] She decided, however, that the nature of the cortex was actually an insufficient criterion. In her 1924 report she said, "As pointed out by Seward, this type of cortex is not confined to a single genus of plants, nor even to a single group, since it is found in pteridosperms and also occurs in some lycopodiaceous stems, and therefore cannot be considered a safe criterion of botanical affinity."[73]

Goldring devoted herself so entirely to her work and to her beliefs and principles that she suffered periodic bouts of exhaustion. One of these periods occurred during 1922 after her intensive work on the Gilboa flora. Concerned for her welfare, Museum director John M. Clarke suggested a trip to his favorite vacation place in the Gaspé area of eastern Quebec for a rest from the museum and to collect fossils. Goldring was hesitant and noted, "Sometimes, when I am feeling extra tired, it seems as though I have not energy enough to even make the trip..."[74] She finally conceded, however, and from July 5 to September 6, 1922, she explored the coast of

Fig. 16. Riverside Quarry along Schoharie Creek where most of the fossil stumps were found, 1920s.

III. GILBOA AND THE FIRST FOSSIL DISCOVERIES

Fig. 17. Fossil stump being removed from Riverside Quarry, 1920s.

Fig. 18. A period of low water in Gilboa Reservoir reveals gray sandstone (Moscow Formation) near Manorkill Falls where impressions of fossil stumps may still be seen, 1991. (photo by the author)

III. GILBOA AND THE FIRST FOSSIL DISCOVERIES

Fig. 19. Impression of a fossil tree stump in gray sandstone (Moscow Formation) near Manorkill Falls, 1991. (photo by the author)

Fig. 20. *Eospermatopteris erianus* Goldring showing what she interpreted as paral-lel strands of the outer cortex. Specimen is 1 meter high, 1 meter (38 inches) wide.

Fig. 21. *Eospermatopteris textilis* Goldring showing what she interpreted as anastomosing strands of the outer cortex. Specimen is 89 cm (35 inches) high, 60 cm (24 inches) wide at base.

Fig. 22. Fossil slab showing the branch systems associated by Goldring with *Eospermatopteris*. These branches are now known as *Aneurophyton*. Specimen is 91 cm (36 inches) long, 62 cm (25 inches) wide

III. GILBOA AND THE FIRST FOSSIL DISCOVERIES

Fig. 23. Restoration of *Eospermatopteris* by Winifred Goldring.

Gaspé Bay and areas to the south, collected fossils, and sent detailed section drawings and field notes back to Clarke.

Revitalized from the trip, Goldring, with assistance from Clarke, planned an exhibit for the Gilboa material now that it had been described in a scientific publication. Many of the stumps had been on display for several months, but it was felt that an exceptional exhibit was required to tell the public about this part of New York's ancient history. With the help of museum artist and sculptor Jules Henri Marchand, Goldring and Clarke conceived an ambitious plan to recreate the Gilboa forests within the museum.

Jules Henri Marchand and two of his sons, Paul and George, operated a "Natural History Studio and Laboratory" in East Schodack, New York. Here they made "models, casts, and restorations of fungi, fruits, flowers, plants, fossils, insects, reptiles and all other natural history accessories for museums and scientific requirements."[75] They were employed by many museums both in New York state and elsewhere during the first half of the twentieth-century as creators of exhibits of excellent quality and imagination. They were instrumental in helping to make the New York State Museum one of the premier natural history museums in the country. The Marchands' exhibits were on display in the State Museum from 1912 until the Museum was relocated from the old State Education Building to the modern Cultural Education Center in 1976. Restorations of Devonian fish, trilobites, a giant beaver, the "Naples Tree," a reproduction of the glacial pothole in which the Cohoes mastodon was found, a mushroom exhibit, and a number of ancient sea bottoms with invertebrates were all important, visually striking exhibits completed by the Marchands for the "old museum." However, the crowning achievement of the Marchands' State Museum work was the Gilboa forest restoration (Fig. 24).

Under the supervision of Winifred Goldring, construction of the Gilboa forest exhibit began in the spring of 1923. The elder Marchand was authorized by the Museum to visit Gilboa to collect any additional material which he felt was needed. In spite of Marchand's brief resignation over salary problems on December 15, 1923 (after which Clarke began paying him out of his own pocket)[76], and numerous technical difficulties posed by the weight of the stumps, the restoration was completed in 1924. An appropriation by the state legislature was required to complete the exhibit. The Gilboa forest was installed in the Main Hall opposite the main elevator on the fifth floor of the Education Building. Because of limited space, other exhibits of plant fossils were installed in the Hall of Vertebrates and in the museum's subgallery at the west end of the hall. These latter two exhibits showed illustrations and specimens of fossil stems, foliage, and seeds associated by Goldring with the stumps.

The Gilboa forest restoration, the largest and most complex restoration in the museum, was about 11 meters (36 feet) wide, 7–9 meters (25–30 feet) high and 4.5–5.5 meters (16–18 feet) deep. The innovative and striking effect of the exhibit featured the recreation of a life-like setting in the foreground that included running water. In addition, the three levels of the forests known in the rock sequence near Gilboa were reconstructed with fifteen fossil stumps positioned on rock ledges against a painted background. Goldring described the intent of the exhibit:

> "In the center foreground flows the Schoharie Creek, which is joined at the left in a series of falls by a tributary, such as the Manorkill. Looking across and beyond this fossil section one sees the painting of our vision of this ancient forest as it might have looked in the height of its glory. The lycopod-like trees (*Protolepidodendron*) [sic], which grew in small numbers in these forests, are also shown in the painting. At both sides of the painting are life-size restorations of the Gilboa tree, which merge imperceptibly into the painting. The artist has depicted so understandingly and skillfully the character of the forest with its heavy moist atmosphere that this restoration is at the same time both a scientific reproduction and a beautiful piece of art."[77]

A large fossil stump was positioned in front of the restoration. This specimen was upended to show the long, radiating roots which held the trees in place in the mud. On both sides of the hall were series of stumps that stood just as they had been found in the quarries at Gilboa.

The official opening of the exhibit took place on Thursday, February 12, 1925, a day when the museum was closed to the public in observance of Abraham Lincoln's birthday. Photographs and postcards of the new exhibit had been sent out with invitations to many geologists from around the country. Distinguished people from Albany, the engineers of the Gilboa dam, major museum directors, the Regents of the University of the State of New York, state assemblymen and senators, and the press were all

Fig. 24. Gilboa forest restoration exhibit — completed in 1924.

III. GILBOA AND THE FIRST FOSSIL DISCOVERIES

invited. This select group enjoyed a private viewing from 3:00 to 6:00 P.M. A formal dedication of the restoration was made at this time by John M. Clarke to the late Canadian paleontologist William Dawson, "as a memorial of the admirable service rendered by Sir William to the science of paleobotany, and as a record of his personal association with the original discovery and study of these trees."[78] An acknowledgement of the dedication was later made in the journal *Science* by J. Austen Bancroft, Dawson Professor of Geology at McGill University.

The restoration was opened to the public the following day and was to become both Winifred Goldring's and the New York State Museum's hallmark. Sadly, it was dismantled after the re-opening of the New York State Museum at the Empire State Plaza in 1976. So widespread was its fame, however, that some out-of-state visitors to the museum still look for it.

John M. Clarke died suddenly on May 29, 1925. In the months following the opening of the restoration requests came pouring into the museum to Winifred Goldring for photographs of the exhibit, reprints of her paper, and *Eospermatopteris* stumps. The latter were supplied to the Redpath Museum of McGill University in Montreal which undertook an exhibit of the restoration[79]; to paleobotanist A.C. Noe of the University of Chicago[80]; the British Museum (Natural History)[81]; the Rijksmuseem in Stockholm[82]; the Museum of Natural History of the University of Rochester[83]; Cornell University[84]; the Geological Survey of Canada at Ottawa[85]; Amherst College[86]; the Geological Museum of Trinity College, Dublin, Ireland[87]; and Goldring's alma mater, Wellesley College[88]. Many of these stumps, including those sent overseas, were supplied by the Hugh Nawn Contracting Company directly from the quarry and shipped at the company's expense. Hugh Nawn took great pride and interest in all of the activities concerning the Gilboa fossils. In fact, a bit of a flap developed over Nawn's personal custody and free distribution of the fossils, since by legal right they belonged to the City of New York.[89]

In June 1927, Goldring sent a letter to Sidney K. Clapp of Grand Gorge, an engineer of the Board of Water Supply, and suggested that some of the stumps be left in place by the Nawn contractors at Riverside Quarry. This would allow development of an in situ roadside exhibit. Clapp responded that there seemed to be some hesitancy on the part of the other Board of Water Supply engineers at Gilboa to sanction this. However, he felt that a letter from the newly appointed New York State Museum Director Charles Adams to Chief Engineer Thaddeus Merriman in New York City would assure compliance.[90] By September, Merriman approved a less dramatic and less paleobotanically valuable plan to gather the largest stump specimens "and place them on concrete foundations near the road or gate so as to be readily accessible for those who are interested." [91] A sketch was provided, and other details, including labeling, were left to Goldring. Unfortunately, Riverside Quarry itself was to be filled in.

In the spring of 1928, Goldring visited the site and provided this description of the completed Board of Water Supply exhibit in New York State Museum Bulletin No. 284 of 1929:

> "This exhibit of the Gilboa fossil trees is situated on New York City ground in lower Gilboa near the junction of the Lower Blenheim–Gilboa road with the road from Gilboa to Grand Gorge. Near-by is the filled-in quarry, Riverside Quarry, from which the greatest number of the fossil trees were obtained and where specimens are still available. The group is just within the fence bounding the city property and can be plainly seen from passing cars. A large-lettered label placed close to the fence can be easily read from a car standing in the road. The exhibit proper occupies a space roughly 20 feet by 4 feet and stumps are set in a cement base thick enough to be unaffected by the action of frost. The stumps have been so arranged that the undersides of the bases are visible in some cases. Around the cement base is a narrow grass plot and in back of the group is a semicircle of evergreens which with a few years' growth will form a very effective background. Between the background of evergreens and the group, is a second detailed label for the benefit of those who are more interested." (Fig. 25).

While the original labels had long since disappeared and it was less visible from the road owing to the growth of the trees, the outdoor exhibit remained on its original site for 73 years. During the summer of 2001, after reaching an agreement with the New York City Department of Environmental Protection, the Town of Gilboa moved the stumps across the creek to a site by the Gilboa Town Hall. Subsequently, the Gilboa Historical Society created a new outdoor exhibit which they dedicated to Winifred Goldring. (Fig. 26).

Fig. 25. The outdoor exhibit of fossil stumps at Gilboa, 1928.

Fig. 26. The new outdoor exhibit of fossil stumps at Gilboa, 2001. (photo by the author)

THE GILBOA FOREST GOES NATIONAL

In November 1927, Winifred Goldring received a letter from Charles Robert Knight (1874–1953). Knight is probably the most famous artist who dealt with paintings of ancient life. Early in his career he achieved recognition for his 1894 restoration of the extinct, pig-like mammal *Elotherium* for Dr. Jacob Wortman at the American Museum of Natural History.[92] Collaborating with such giants of paleontology as Edward Drinker Cope, Henry Fairfield Osborn, and Roy Chapman Andrews, Knight produced scores of outstanding restorations and paintings of prehistoric life. In 1926, he was commissioned by the Field Museum in Chicago to create a series of 28 murals to line Graham Hall, their new vertebrate fossil gallery.[93] The purpose of the series was to illustrate the history of life from the Cambrian to the present. Knight had heard about the Gilboa restoration and wanted photographs and suggestions from Winifred Goldring concerning "the character of the foliage"[94] for a Devonian mural that was to be 8 meters (25 feet) long by 2.7 meters (9 feet) high. Goldring was delighted by Knight's interest and responded with a long letter that included detailed descriptions of Devonian flora and fauna, the interpreted nature of the climate, and references to other reports that described Devonian plants.[95]

In April 1928, Knight stopped in Albany to see the restoration on his way from his home in New York City to Chicago. The final result of his collaboration with Winifred Goldring was an evocative painting of a Devonian coastal lowland area beautifully and faithfully rendered with great sensitivity to the suggestions given him. Since its completion in 1930, it has been reproduced many times, as have all the murals of the series, and is on display in the Field Museum's recently renovated Graham Hall. The continued display of the Devonian mural is an enduring tribute to both Knight and Goldring. (Fig. 27).

GILBOA TREES ARE NOT SEED FERNS

The German paleobotanist Richard Krausel (1890–1966) and his associate Herman Weyland studied Devonian plants in Europe. In 1928, Krausel journeyed to the United States and Canada to collect plant fossils at many localities and to use museum collections. He published a paper in 1941 on Devonian plant remains in North America.[96]

Krausel worked on *Eospermatopteris* by comparing material found in New York state with that from similar age rocks in Germany and brought new information to light. In a 1935 report, Krausel and Weyland showed that the distinctive cortical pattern of *Eospermatopteris textilis* was actually part of the deep, inner cortex of *Eospermatopteris erianus*. This meant that the two species were identical and the Gilboa stumps were called *Eospermatopteris erianus*.[97] *Eospermatopteris erianus* was selected as the appropriate name and *E. textilis* was abandoned because *"erianus"* appeared first in Dawson's 1871 description. The German workers then showed that the "seeds" described by Goldring contained spores and, thus, were not seeds at all but *sporangia*.[98] Similarly, Goldring's "microsporangia" were determined to be nothing more than inorganic markings.[99] The important consequence of Krausel and Weyland's work was to demonstrate that the Gilboa trees were not "seed ferns."

The final conclusion reached by Krausel and Weyland was that the branch systems of the Gilboa fossils were like those of the progymnosperm *Aneurophyton*, described by Krausel and Weyland in 1923 from German specimens. The New York branch material then became *Aneurophyton erianum* for a while in the scientific literature. Later, in 1978, paleobotanists Bruce Serlin and Harlan Banks assigned newly collected New York branch fossils to *Aneurophyton germanicum* of Krausel and Weyland. *"Aneurophyton erianum"* was abandoned as a name because reexamination by Serlin and Banks of Goldring's material upon which the name was based showed that she was actually dealing with fossils of more than one type of plant.[100]

At present, the Gilboa stumps are called *Eospermatopteris erianus*. The associated but unattached branch systems are called *Aneurophyton*.[101] Whether the two types of fossils are actually parts of the same plant remains a mystery. More recent interpretations of *Eospermatopteris* suggest that perhaps it was not a large tree at all, but rather a short stubby plant like a cycad.[102]

The "dismantling" of *Eospermatopteris* in this way may seem to place the quality of Goldring's work in question. However, great difficulties are inherent in all restorations of extinct life forms, particularly plants, since they are almost never found in a complete state. None of the individual fossil parts of the Gilboa tree were ever found associated in one fossil, nor has any complete specimen of a Devonian tree with a trunk bearing *Eospermatopteris* characteristics been found anywhere to this day. The discovery of a whole specimen would indeed be a significant event; it would be the only way to determine whose ideas concerning *Eospermatopteris* are actually correct.

Fig. 27. Charles Knight's painting based, in part, on the Gilboa forest restoration exhibit of the New York State Museum. (Used with permission — The Field Museum, #CK73898, Chicago)

OTHER FOSSIL PLANTS FROM GILBOA

Just after Winifred Goldring's trip to Gaspé in 1922, Loren Petry, a botany professor at Cornell University, began collecting fossils along the Gaspé and northern New Brunswick coasts. This area had afforded Dawson many collecting sites for Early to Middle Devonian plants but had seemingly languished in terms of exploration since Dawson's investigation in the mid-1800s. Petry made extensive collections of Devonian plant fossils which he subsequently reposited at Cornell.[103] While not a well-published paleobotanist, Petry did convey his interest and enthusiasm in paleobotany to some of his students at Cornell, among them Harlan Parker Banks (1913–1998), who continued work on Devonian plants in southeastern Canada and northern Maine. Banks, in particular, was to become one of the preeminent Devonian plant workers in the United States and eventually focused his efforts on the Devonian flora of New York state. His work and that of his students included the introduction of new and modified laboratory and collecting techniques, and numerous scientific publications which revived an interest in Devonian plant research in the United States. Many of the fossil plants described by Banks and his students were based on specimens from Gilboa.

In 1937, Chester A. Arnold, another former student of Loren Petry, illustrated herbaceous lycopod fossils that consisted of "several impressions of spiny stems" which he had collected from Riverside Quarry in 1926 and 1929. He erected a new genus and species, *Gilboaphyton goldringiae*, in honor of Winifred Goldring[104] (Fig. 28). However, the German paleobotanists Krausel and Weyland (1949) reinterpreted this plant as simply another specimen of a known herbaceous lycopod, *Archaeosigillaria vanuxemii*. Their reasoning was based on the observed presence of a hexagonal pattern of leaf scars on a portion of the stem, a characteristic of *Archaeosigillaria*. In 1963, Harlan Banks and his graduate student James D. Grierson published a comprehensive study, *Lycopods of the Devonian of New York*, which provided a complete review of the known fossil lycopods of New York. After reexamining the type material of *A. vanuxemii* and examining hundreds of other specimens in the Cornell University collections and their own collections from Gilboa, they concurred with Krausel and Weyland that *Gilboaphyton goldringiae* is a junior synonym of *Archaeosigillaria vanuxemii*.[105]

Later, in 1978, Banks and Muriel Fairon-Demeret, a Belgian paleobotany student, applied the painstaking "degagement" technique on

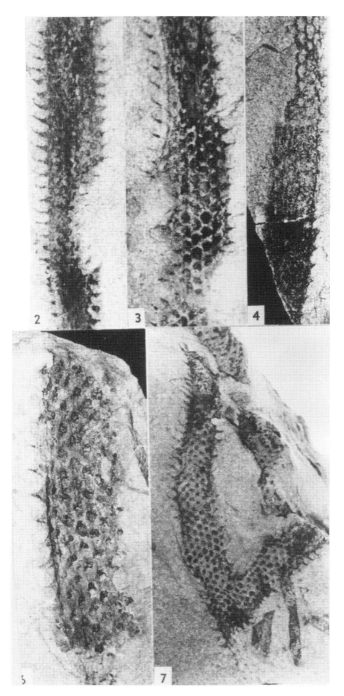

Fig. 28. Specimens of *Gilboaphyton goldringiae* from Riverside Quarry at Gilboa showing, in particular, the hexagonal pattern of leaf scars on the stems, (x 1.2, x 1.2, x 1.4, x 1.4, x 0.9). (courtesy of Dr. Harlan P. Banks)

Archaeosigillaria vanuxemii. The degagement technique was developed by Suzanne Leclercq in 1960 as a means to recover plant fossil material. It involves the use of a small hammer and needles and a low-power binocular dissecting microscope to remove plant material from its rock matrix. This technique confirmed both the flattened nature of the leaves of *A. vanuxemii* and showed their toothed margins and terminal "hair."[106]

Most recently (1997), Christopher Berry and Dianne Edwards reinterpreted the details of the stem anatomy of *Archaeosigillaria vanuxemii.* They reassigned these plants from Gilboa to the genus *Gilboaphyton.*[107]

In their 1963 study, Grierson and Banks also included some specimens that Goldring believed had impressions of petiolar scars. "Petiolar" scars are the impressions of the bases of leaves that have been shed. These specimens were brought by Grierson and Banks to their new genus *Amphidoxodendron* ("doubtful tree"). This genus includes arborescent lycopods that have large, spirally arranged leaf scars on their stems.

Sigillaria? gilboensis, a lycopod described by Goldring in 1926 from poorly preserved specimens collected by a Hugh Nawn Company engineer at the Gilboa dam site, was also reexamined by Grierson and Banks. Like Goldring, they could not assign the specimens definitely to *Sigillaria* because of poor preservation quality. These specimens remain in question until better material can be found (Fig. 29).

Other herbaceous lycopods described from Gilboa during the 1960s by Harlan Banks included *Protolepidodendron scharianum,* a species based on specimens found at the mouth of Manorkill Creek in 1922 by Winifred Goldring's assistant Joseph Bylancik, and *Protolepidodendron gilboense,* based on specimens from Riverside Quarry. The specimens of *P. scharianum* represent the only known occurrence of this lycopod in the United States, and *P. gilboense* is known only from Gilboa[108] (Fig. 30).

A fossil found in a spoil heap along Schoharie Creek at Gilboa in the 1920s and placed in the collections of the New York State Museum was reexamined and identified as *Pseudosporochnus verticillatus* in 1966 by Penny G. Schuchman, another of Banks' graduate students. This plant, probably a shrub or small tree, belonged to a group of plants known as the Cladoxylales. These plants were unique because they had a number of characteristics which, individually, were particular to several different plants. However, the Cladoxylales became extinct at the end of the Early Carboniferous and left no descendants. The *Pseudosporochnus verticillatus* specimen from Gilboa is especially significant because, according to

Fig. 29. *Sigillaria? gilboensis*. Specimen is 81 cm (32 inches) long, 30 cm (12 inches) wide.

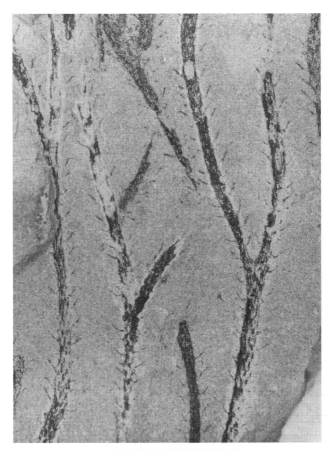

Fig. 30. *Protolepidodendron gilboense.* (Courtesy of Dr. Harlan P. Banks)

Fig. 31. Reconstruction of *Lepidosigillaria* ("Naples tree") in the "old" New York State Museum.

Schuchman, "it is the first specimen of the genus from North America on which leaves have been found in attachment. It is also significant because it is the only specimen of the species from which anatomy has been described. Previous determinations have been based solely on external features."[109]

In 1973, a new genus of Devonian plant was discovered by graduate student J. E. Skog and Harlan Banks in a large block found near the site of Riverside Quarry. This plant, named *Ibyka amphikoma*, had a large branch system. As in the case of *Aneurophyton, Ibyka's* presence in the same rock sequence as *Eospermatopteris* makes it another possible candidate for being the foliage of the *Eospermatopteris* stumps.[110]

Two other plants found at Riverside Quarry were *Prosseria*, a "horse-tail," and the arborescent lycopod *Lepidosigillaria. Lepidosigillaria* is one of the most interesting Devonian plants that has been found in New York state. It has been called the "Naples tree" because a large specimen was found in black shale near the western New York village of Naples in 1882. The black shales were originally deposited as black mud in relatively deep sea water. The "Naples tree" originally grew in a forest on the Catskill Delta but was uprooted and transported to sea by rivers that crossed the delta. After floating west, the tree became waterlogged, sank to the sea bottom, and was buried in mud. The trunk of the "Naples tree" is in the New York State Museum. It measures about 3.6 meters (twelve feet) in length and the surface is covered with petiolar scars. Some researchers believe that *Lepidosigillaria* may actually be the trunk portion of the enigmatic *Eospermatopteris* stump.[111]

Chapter IV

A New Quarry, New Plants,
and Discovery of the First Land Animals
at Gilboa

It's important to live life with a knowledge of its mystery and of your own mystery.

– Joseph Campbell

Water and the history of the Gilboa area are inextricably linked. Once the region was part of the huge Catskill Delta complex. Some 380 million years later, the availability of water from Schoharie Creek for mills, agriculture, and domestic purposes drew settlers to the area. A 19-century flood of the creek severely damaged the settlement and indirectly brought about an important fossil discovery. Then, the demand by New York City for water from this portion of the watershed destroyed the village of Gilboa, while at the same time prompting many important fossil plant discoveries.

In 1968, the New York Power Authority was granted permission from the state legislature to build a pumped storage hydroelectric facility on Schoharie Creek. This facility was planned for an area north of the Gilboa dam and reservoir that encompassed hundreds of acres of land in the Town of Gilboa and the Town of Blenheim.[1] A federal license was issued in 1969 and construction of the Blenheim-Gilboa Pumped Storage Facility began with ground-breaking ceremonies in July 1969.[2] Residents of the area were compelled to sell farm land and, in a few cases, their homes. The flow of Schoharie Creek was further impacted, and the effects were felt by many downstream. On the positive side, construction exposed large sections of fossil-bearing rock.

The power facility operates by pumping water from a large, lower reservoir up to a smaller, higher one. This water is then released and rapidly descends 1000 feet to the turbine generators below.[3] The higher reservoir was built on the top of Brown Mountain east of Schoharie Creek in the Town of Gilboa. The generating plant lies on a platform that faces a cliff cut into the west flank of Brown Mountain. Raymond A. Baschnagel, a high school biology teacher from Delhi, New York, examined this cliff for Harlan Banks at Cornell and Patricia M. Bonamo and James D. Grierson at State University of New York at Binghamton. Baschnagel soon discovered a lens of rock with plant fossils that was exposed during blasting. This lens was a paleobotanist's dream-come-true because it represented a pond, lake, or stream-side deposit that was filled with profuse vegetation. Subsequent collection and examination of this late Middle Devonian (Panther Mountain Formation) material by Bonamo and Grierson showed it to be a fine-grained mudstone literally packed with long (up to 46 cm), slender, trailing, leaf-covered lycopod stems (Fig. 32). The stems had formed thick mats close to the soil. Each leaf was forked into five parts. By using a technique known as "maceration," which involves dissolving rock in hydrofluoric acid, and a bioplastic transfer method developed by Banks, Bonamo, and Grierson, the actual remains of the plants were retrieved in the form of black (carbon) compressions and pyrite petrifactions.[4] In an effort to reveal details in the compressions, Bonamo and Grierson tried several different oxidation techniques to remove the carbon and pyrite. This made the fossils almost transparent and allowed close examination of internal structure.[5] The preservation was extraordinary. It could be seen that each leaf had one vein and stomata or pores. Sporangia (spore sacs) still filled with spores were found on the surfaces of some of the leaves (Fig. 34, 35). Xylem tissue preseved in some of the pyritized stems revealed fine details of anatomy when polished. Even sugar-filled phloem tissue, the first tissue generally destroyed by decay-producing bacteria, was preserved in some of the stems.[6] Described in 1972 by Banks, Bonamo, and Grierson as a new genus of lycopod, the plant was named *Leclercqia complexa* in honor of Belgian paleobotanist Suzanne Leclercq. The species name refers to the distinguishing characteristic of the plant — the complex, five-fold division of its leaves (Fig. 33).

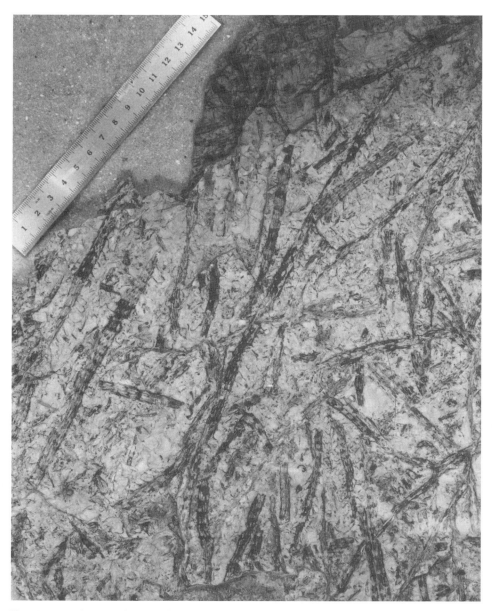

Fig. 32. Mudstone slab packed with *Leclercqia complexa* axes or stems, (x 0.6). (courtesy of Dr. Patricia M. Bonamo)

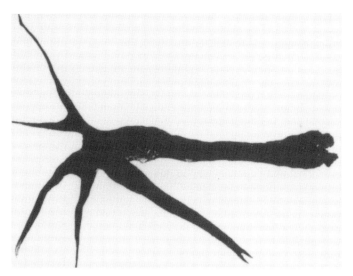

Fig. 33. *Leclercqia complexa* leaf macerated out of its rock matrix, (x 18). (courtesy of Dr. Patricia M. Bonamo)

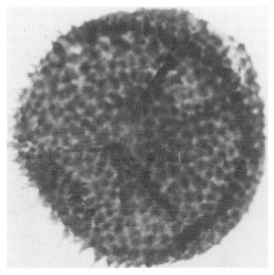

Fig. 34. Spore of *Leclercqia complexa*, (x 920). (courtesy of Dr. Patricia M. Bonamo)

Fig. 35. Sporangium of *Leclercqia complexa* containing spores, (x 42). (courtesy of Dr. Patricia M. Bonamo)

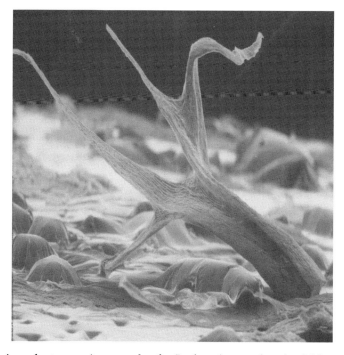

Fig. 36. Scanning electron micrograph of a *Leclercqia complexa* leaf. Note the ligule just above the base of the leaf, (x 40). (courtesy of Dr. Patricia M. Bonamo)

Further examination of the leaves by Grierson and Bonamo with a scanning electron microscope revealed the presence of a *ligule*, a kind of appendage on the surface of some of *Leclercqia's* leaves. The presence of this enigmatic organ on *Leclercqia* was totally unexpected and added to its mystery. For example, the presence of a ligule in extant lycopods is confined to heterosporous plants (those having microspores that grow into male gametophytes and megaspores that grow into female gametophytes); *Leclercqia* was, presumably, homosporous (having monomorphic spores). *Leclercqia's* ligule is located at some distance from the attachment of the leaf to the stem, a position that differs from any known ligulate lycopod. Interestingly, the morphology of *Leclercqia's* ligule is the same as the ligules of extant lycopods, a fact that unfortunately sheds no new light on its function.[7] *Leclercqia* is the earliest known homosporous lycopod with a ligule, and the ligule on *Leclercqia* is the oldest known ligule in the fossil record[8] (Fig. 36).

The combination of superb preservation and painstaking and imaginative preparation on the part of Banks, Bonamo, and Grierson has made *Leclercqia* the best illustrated extinct herbaceous lycopod. The observations and interpretations of extinct lycopods based on *Leclercqia* reinforce the similarities between extinct and living types. In nearly 400 million years this plant group has changed very little. No other group of vascular land plants has a continuous record as long as the lycopods. If no subsequent new material had ever been recovered from Brown Mountain near Gilboa, the site would still be widely known and appreciated because of the importance of the information gained from *Leclercqia* to our understanding of this ancient plant line.

Collections made at Brown Mountain during 1971 and 1972 revealed another type of plant known as *Rellimia thomsonii* (Fig. 37). Compressions, impressions, and pyrite petrifactions of these plants were also beautifully preserved. The somewhat fragmentary nature of the specimens implied that the depositional site of these plants was not the original growth site as was the case for the nearly complete remains of *Leclercqia*. However, the growth site was probably not far away.

Rellimia thomsonii, a progymnosperm (i.e., precursor of the seed plants), was known from specimens in Europe and Russia, but discovery at Gilboa was the first record of this plant in North America, and Patricia Bonamo's was the first description of its anatomy. Its record in the New York portion of the Catskill Delta, along with such similar plants as

Fig. 37. Branch system of *Rellimia thomsonii* in rock matrix, (x 0.49). (courtesy of Dr. Patricia M. Bonamo)

Aneurophyton and *Tetraxylopteris*, raises questions concerning the evolutionary relationships of these plants since they share some major morphological characteristics.[9]

A later study of *Rellimia thomsonii* from Gilboa by Dannenhoffer and Bonamo demonstrated the existence of growth layers in the wood of this

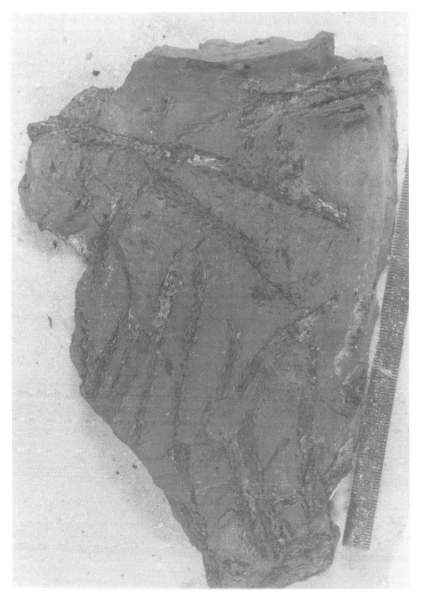

Fig. 38. *Haskinsia* in rock matrix. (courtesy of Dr. Patricia M. Bonamo)

plant. This was "the first clear documentation of gymnospermous wood with growth layers in the Middle Devonian plants."[10] This observation provided evidence of seasonality, perhaps alternating wet and dry climates for this locality in the Middle Devonian. An absence of growth layers implies a seasonless climate; marked layers are evidence of seasonality.

On the basis of specimens from the Gilboa site, yet another genus of lycopods was described in 1983 by Grierson and Banks. The genus name *Haskinsia* was given to these plants in honor of the late Vernon Haskins of Durham, New York, a naturalist and local historian who directed paleobotanists to many plant fossil localities in the Catskills[11] (Fig. 38).

The fine preservation of the Gilboa material is largely due to rapid burial in fine-grained sediment and an anoxic burial environment which inhibited decay. Another factor was the proximity of the depositional site to the material's original growth site; this meant that plant material was not damaged by long-distance transport by streams. In working with lycopods, this high-quality preservation has allowed researchers to understand the importance of leaves in evaluating the relationships between poorly or incompletely preserved specimens.[12] Most Devonian lycopods are basically so similar in gross morphology and habit that only exceptional preservation, like that at Gilboa, can reveal subtle differences in their small leaves and permit these leaves to be used as diagnostic characters. For example, the leaves of *Leclercqia* fork into five segments with a middle, sharply recurved segment. The leaves of *Haskinsia*, in contrast, are simple and taper from a thickened base to an acute, unforked tip. *Protolepidodendron* leaves have an enlarged base, but the tips are forked into two parts. *Leclercqia*, *Haskinsia*, and *Protolepidodendron* are so similar in gross appearance that in cases of poor preservation or in the absence of leaves it is virtually impossible to tell them apart.

In addition to the higher plants discussed above, fossil evidence of filamentous algae and fungi has also been found at Gilboa.

DISCOVERY OF DEVONIAN LAND ANIMALS

While examining some of the macerated *Leclercqia* material in her lab at SUNY Binghamton, paleobotanist Patricia Bonamo found what appeared to be tiny animal legs with the plant material. Later, her colleague at Binghamton, James Grierson, noticed a tiny, flattened, spider-like animal attached to one of the plant samples. He recovered this animal and the fragments of several others, mounted them on slides, and set them aside for further study. Bonamo and Grierson recovered the cuticles (outer coverings made of chitin) of more fossil arthropods as their work progressed. They now had the first examples of tiny terrestrial animals that were a part of the 380 million-year-old plant community their research was trying to reconstruct.[13] They knew this, but getting other people to believe it was very nearly impossible! The paleozoologists who they approached were all skeptical. Ironically, the extraordinary preservation of the fossils was now a liability! The cuticles of the animals seemed to be too well preserved (even tiny hairs were visible). Even though these remains were fragmentary, had pyrite pits in them, and were compressed like the plants, they were regarded simply as contemporary animals by other scientists.[14] They were believed to be contaminants that fell from the light fixtures in the lab or emerged from cracks in the walls regardless of the precautions taken by Bonamo and Grierson to prevent such contamination. Finally, ten years later, Ian Rolfe, Keeper of Geology at the Royal Museum of Scotland, visited Bonamo and Grierson in 1981 to look at the animal remains. He came away with some photographs and the impression that the animals were indeed old. But when he met zoologist William A. Shear at a meeting in Virginia shortly after this visit and showed him the photographs, the two of them became convinced of the great age of the Gilboa remains. Shear's expertise in living arthropods allowed him to recognize the relationship of the Gilboa fossils to other long-extinct animals. Bonamo and Grierson agreed to let Shear collaborate with them, and they organized an international team of experts to describe the animals and their paleoenvironment. The Gilboa fossils actually changed Shear's career; he became so intrigued that he subsequently changed the focus of his research from modern soil arthropods to fossil arthropods and the reconstruction of early terrestrial ecosystems.[15]

Much more animal material was extracted from the Gilboa rock, and the team of researchers slowly began to fit the pieces together. "The animal fossils are preserved as minute, undistinguished brown to black flakes, which are unrecognizable as animals when in the rock under incident light microscopy, but transmitted light reveals their zoological nature."[16] Sorting and reconstructing the fragments revealed a variety of animals with a "striking predominance of predatory arthropods."[17]

In 1987, three new genera and seven new species of the arachnid order Trigonotarbida were described by Shear, Paul Selden of England's

Manchester University, Ian Rolfe, Bonamo, and Grierson. These extinct relatives of spiders have armored plates on their abdomens and lack silk-producing organs.[18] In contrast to spiders which have "simple" eyes consisting of a single lens, trigonotarbids have compound eyes with ten or twelve lenses.[19] They are the most abundant animal fossils found at Gilboa and at other terrestrial sites such as the older Rhynie Chert of Scotland.[20]

One of the three new trigonotarbid genera, *Gilboarachne*, commemorates Gilboa. Among the seven species one is named *Gelasinotarbus bonamoae*, and another is *Gilboarachne griersonii* to honor the discoverers of the Gilboa animals (Figs. 39, 40).

Centipedes are small, terrestrial, predatory arthropods whose fragile exoskeletons and mode of life do not normally make them likely candidates for fossilization. Yet at Gilboa, more than thirty specimens (predominantly heads and poison-claws) have been recovered. These represent a type of small (ca. 10 mm) centipede that is assigned to a new order of these animals — the Devonobiomorpha — based on their unique poison-claws.[21] The species *Devonobius delta* (refers to the Catskill Delta) is among the oldest known centipedes[22] (Figs. 41, 42). It is believed that the mats of *Leclercqia complexa* stems and leaves, within which *Devonobius delta* and the other animals have been found, acted as a sieve that trapped the dead bodies of the animals. Due to this post-mortem entrapment it is difficult to determine whether the animals actually lived among these particular plants.[23]

Perhaps one of the most interesting arthropod finds from Gilboa was that of a nearly complete spider spinneret (Fig. 43). This organ produces silk for webs, egg sacs, and other uses in modern spiders. According to Shear and spinneret experts Jackie Palmer of Western Carolina University and Jon Coddington of the Smithsonian, the Gilboa spinneret represents the earliest known evidence of silk production and hence, the earliest record of spiders.[24] It is unclear to what uses Devonian spiders put their silk. Modern spiders with this particular type of spinneret (the mesotheles) use silk for lining their burrows, for constructing trip lines from the burrows to warn of approaching prey, and for making egg sacs. However, they do not make webs.[25] The Devonian spiders may also have used their silk in this way, because flying prey for which most webs are constructed would not appear for several million years.

Fossil spider expert Paul Selden was later able to put the spinneret together with other fragments, including jaws and poison-fangs, to reconstruct the actual animal. This is *Attercopus* (poison spider) — the oldest known spider.

Another type of arachnid, *Ecchosis pulchribothrium*, was also described from the Gilboa fossils by Selden, Shear, and Bonamo in 1991.[26]

A very strange looking animal was discovered by Bonamo and nicknamed the "Angry Dragon" because of its lethal looking open jaws. This fossil turned out to be that of a pseudoscorpion. These tiny, scorpion-like animals that lack stingers today live as predators in the soil. The fossil was studied by fossil pseudoscorpion expert Wolfgang Schawaller of the Stuttgart Museum in Germany. This fossil and a second specimen were eventually named *Dracochela* or "dragon pincer."[27] *Dracochela* is 300 million years older than any previously described pseudoscorpion. The spinnerets found on the animals provide the oldest evidence of silk production in pseudoscorpions[28] (Fig. 44).

Fig. 39. Trigonotarbid *Gilboarachne griersonii* — *greatly* magnified. (courtesy of Dr. Patricia M. Bonamo)

Fig. 40. Trigonotarbid *Gelasinotarbus bonamoae* — *greatly* magnified. (courtesy of Dr. Patricia M. Bonamo)

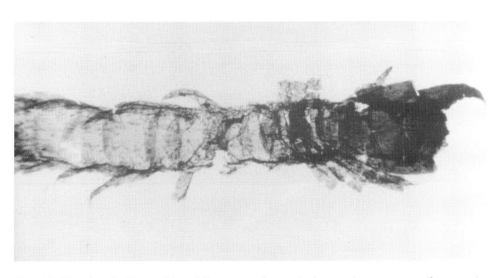

Fig. 41. Centipede *Devonobius delta* — nearly a whole specimen — *greatly* magnified. (courtesy of Dr. Patricia M. Bonamo)

Fig. 42. Centipede *Devonobius delta* — head and maxillipeds, partially connected — *greatly* magnified. (courtesy of Dr. Patricia M. Bonamo)

Fig. 43. Spinneret of the oldest spider *Attercopus fimbriunguis* — *greatly* magnified. (courtesy of Dr. William A. Shear)

82

IV. A NEW QUARRY, NEW PLANTS, AND
DISCOVERY OF THE FIRST LAND ANIMALS AT GILBOA

Fourteen whole or partial mite specimens were recovered from the Gilboa material. These fossils were described by John B. Kethley of Chicago's Field Museum, Roy A. Norton of the SUNY College of Environmental Science and Forestry at Syracuse, Bonamo, Grierson, and Shear. New genera and species of mites include *Protochthonius gilboa*, *Devonacarus sellnicki*,[29] and *Archaeacarus dubinini* (Figs. 45–47). The discovery of the latter species at Gilboa helps to substantiate the previously questioned authenticity of similar animals found in the Rhynie Chert.[30]

Arthropleurids were found in abundance in the Gilboa material. While these millipede-like animals were probably *detrivores* (animals that live on dead plant material), they likely served as prey for the spiders, trigonotarbids, and centipedes.[31]

Archaeognaths, a group of wingless insects known as bristletails, were also found among the fragments of cuticle.[32]

While the Gilboa fauna includes the oldest known representatives of several types of terrestrial animals, the very earliest known terrestrial arthropods (along with the tiny vascular plant *Cooksonia*) were reported in 1990 from Upper Silurian rocks in Shropshire, England.[33]

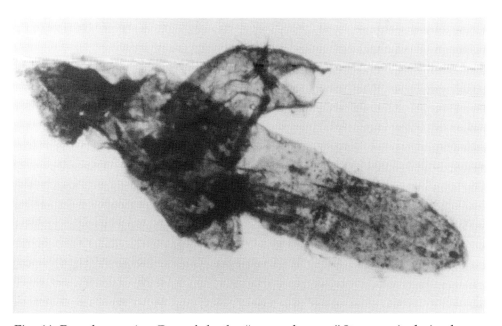

Fig. 44. Pseudoscorpion *Dracochela*, the "angry dragon." Its name is derived from its "jaws" at the center of the photo. *Greatly* magnified. (courtesy of Dr. Patricia M. Bonamo)

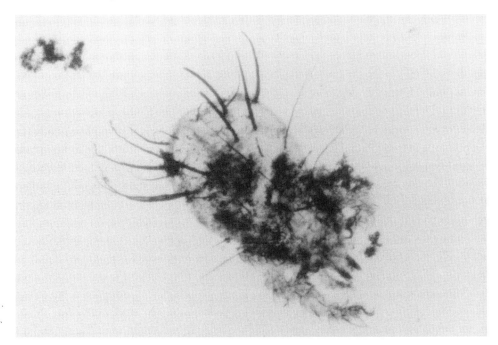

Fig. 45. Mite *Protochthonius gilboa* — *greatly* magnified. (courtesy of Dr. Patricia M. Bonamo)

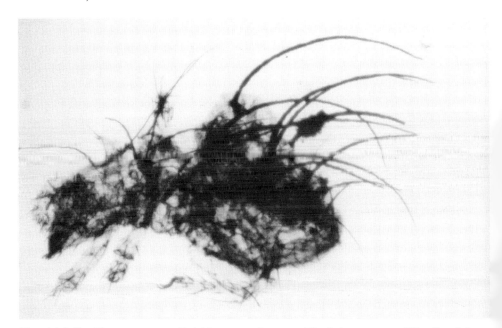

Fig. 46. Mite *Devonacarus sellnicki* — *greatly* magnified. (courtesy of Dr. Patricia M. Bonamo)

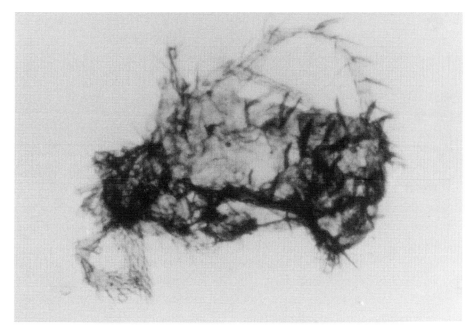

Fig. 47. Mite *Archaeacarus dubinini* — *greatly* magnified. (courtesy of Dr. Patricia M. Bonamo)

The effectiveness of techniques used in the Gilboa research (i.e., the breakdown of rock samples by hydrofluoric acid), in conjunction with the evidence from Gilboa for the existence of a terrestrial ecosystem with well-differentiated niches in the Middle Devonian has prompted a revolution in paleontology. The Gilboa discoveries have induced some researchers to reexamine Devonian fossil material from other areas for the presence of arthropods, and propelled others, such as those in England, to search in ever older rocks for the onset of the "land invasion."

GILBOA IN THE DEVONIAN — ITS IMPORTANCE AS A SETTING

When we speak of our heritage it is nearly always in terms of the past two or three hundred years, particularly in the United States where it is difficult to trace family lineage beyond just a few generations. But it is important at times to consider the broader view of our heritage and to include a vision of "deep time" that encompasses animals that lived much before the age of dinosaurs. This heritage includes the Middle Devonian Gilboa fossils that extended back some 380 million years and lived along the shoreline of the inland Catskill Sea. This was a tropical world, situated about twenty degrees south of the paleoequator.[34] We can reconstruct this world in considerable detail: Highlands composed of uplifted Paleozoic sedimentary rock lay far in the distance to the east.[35] Arborescent or tree-like plants — *Aneurophyton, Lepidosigillaria*, and *Eospermatopteris* — grew in abundance a little distance from the water's edge and competed for space with such large, shrubby plants as *Rellimia, Ibyka, Pseudosporochnus*, and horsetails such as *Prosseria*. Dense mats of lycopods — *Haskinsia, Leclercqia, Archaeosigillaria*, and *Protolepidodendron* — thrived, particularly on stream banks and around ponds. In the soil and among the plants, a diverse assortment of arthropods moved about — detrivores and their predators. Plants suffered no damage from herbivores for insects and other animals with a herbivorous habit had not yet developed. Air-breathing rhipidistian fish came ashore from time to time, possibly in search of food or to find a more adequate freshwater environment. The air, when not stirred by wind, was languid, undisturbed by wings or the sounds of animal songs or cries. Long periods of desert dryness were followed by monsoon rains and flooding. Overflow from flooded streams often buried lycopod fragments in fine mud in flood plain deposits along with some of the arthropods which had been washed into the mats. Another occurrence of plant remains and arthropods is in poorly oxygenated deeper lake beds at the front of small deltas. Occasionally (perhaps every 500 or 1000 years), a storm of extreme intensity produced waves that drove the barrier islands landward. The barrier islands were long, narrow accumulations of sand lying a few miles offshore and oriented parallel to the shore. The result was the flooding and immediate burial of much of the forest in sand which killed the plants. In time, the trunks and foliage of trees like *Eospermatopteris* rotted, disintegrated, and floated away. The molds of the stumps left in the sand were eventually filled in by more sand, the result of a rise in sea-level.

Gilboa is a window on the history of the land colonization process, a pivotal, world-wide process laden with possibility but at the same time fragile in its beginning and fraught with contingencies. While all windows on the past are important for bringing us closer to the reality of evolution, Gilboa is perhaps more relevant for us in that here we may see and appreciate the sort of setting in which our *real* ancestors — amphibious, fish-like tetrapods — breathed their first air and took their first steps.

Research on the Gilboa fossils continues…

NOTES AND REFERENCES

Chapter I

1. Banks, H.P. 1970. Evolution and Plants of the Past. Wadsworth Publishing Co., Inc., Belmont, CA., p. 22.

2. Thomas, B.A., and R.A. Spicer. 1987. The Evolution and Paleobiology of Land Plants. Dioscorides Press, Portland, OR., p. 6.

3. Edwards, D., M.G. Bassett, and E.C.W. Rogerson. 1979. The earliest vaocular land plants: continuing the search for proof. Lethaia, 12:313–324.

4. Wright, V.P. 1990. Terrestrialization, p. 57. In D.E.G. Briggs and P.R. Crowther, eds., Palaeobiology: A Synthesis. Blackwell Scientific Publications, Boston.

5. Ibid.

6. Ibid.

7. Ibid, p. 58.

8. Thomas, B.A., and R.A. Spicer. 1987. The Evolution and Palaeobiology of Land Plants. Dioscorides Press, Portland, OR., p. 13.

9. Shear, W.A. 1991. The early development of terrestrial ecosystems. Nature, 351: 283–289.

10. Blatt, H., G. Middleton, and R. Murray. 1980. Origin of Sedimentary Rocks. Prentice-Hall, Inc., Englewood Cliffs, N.J., p. 29.

Chapter II

1. Rudwick, M.J.S. 1985. The Great Devonian Controversy. The University of Chicago Press, Chicago and London.

2. Dinely, D.L. 1984. Aspects of a Stratigraphic System: The Devonian. John Wiley and Sons, New York, p. 186.

3. Ibid.

4. Ibid.

5. Ibid., p. 142.

6. Ibid., p. 187.

7. Algeo, T.J., S.E. Scheckler, and J.B. Maynard. 2001. Effects of the Middle to Late Devonian spread of vascular land plants on weathering regimes, marine biotas, and global climate. *In* P.G. Gensel and D. Edwards, eds., Plants Invade the Land. Columbia University Press, New York, pp. 213–236.

Chapter III

1. Near Middleburgh in Schoharie County, two streams enter the Schoharie Creek in such a way that a back current is created which causes driftwood to accumulate. Local lore has it that "Schoharie" is derived from the Iroquois word for driftwood as observed in this area.

2. Dineen, R.J. 1987. Schoharie Creek flood of April 5, 1987 — A preliminary report. N.Y.S. Geological Survey Open File Report, Albany, p.2.

3. Also called "Plattenkill" according to Roscoe, W.E. 1882. History of Schoharie County. D. Mason and Co., Syracuse, p. 121.

4. Ibid.

5. Dictionary of scripture proper names. 1983. *In* Websters New Universal Unabridged Dictionary, p. 90. Simon and Schuster, New York.

6. Roscoe, W.E. 1882. History of Schoharie County. D. Mason and Co., Syracuse, p. 122.

7. Fanning, B. History of Gilboa, New York from 1839–1861. Published by the author, p. 29.

8. Ibid.

9. Baily, M. 1975. Floods and dams on Schoharie Creek at Gilboa. *In* M.V.O. Norton, ed., Schoharie County Historical Review, p.7.

10. Hendrix, L.E. and A.W. 1988. Sloughter's Instant History of Schoharie County, p. 103.

11. Ibid.

12. Ellis, F. 1885. History of Monmouth County, New Jersey. (Publisher unknown), p. 446.

13. Sketch of the Reformed Church of Keyport, New Jersey. 1926. (Publisher unknown).

14. Ellis, F. 1885. History of Monmouth County, New Jersey. (Publisher unknown), p. 447.

15. Cope, E.D. 1870. Extinct Batrachia, Reptilia and Aves of North America. Transactions of the American Philosophical Society, v. 14, pp. 121, 122.

16. Ibid., p. 40.

17. Ibid., p. 127.

18. Hall, J. 1871. Report of the Director, Twenty-fourth Annual Report on the State Museum of Natural History, The Argus Company, Albany p. 16.

19. Ellis, F. 1885. History of Monmouth County, New Jersey, (Publisher unknown), p. 446.

20. Andrews, H.N. 1980. The Fossil Hunters. Cornell University Press, Ithaca, p. 195.

21. Dawson, J.W. 1871. The Fossil Plants of the Devonian and Upper Silurian Formations of Canada. Dawson Brothers, Montreal, p. 59.

22. Dineen, R.J. 1987. Schoharie Creek flood of April 5, 1987 — A preliminary report. N.Y.S. Geological Survey Open File Report, Albany, p. 5.

23. Baily, M. 1975. Floods and dams on Schoharie Creek at Gilboa, *In* M.V.O. Norton, ed., Schoharie County Historical Review, p. 6.

24. Proceedings of the Albany Institute from March 1865 to September 1872, Volume 1. Albany, NY., J. Munsell, 1873, p. 130

25. Letter of D. Mackey of Gilboa to "Mr. Baker" requesting that he call upon James Hall to inform him of the discovery of stumps at Gilboa, January 3, 1870 New York State Archives, Albany, NY.

26 Records of the Reformed Church of Gilboa, New York.

27. Hall, J. 1871. Report of the Director, Twenty-fourth Annual Report of the Museum of Natural History, The Argus Company, Albany, p. 16.

28. Ibid., p. 21.

29. Ibid., p. 8.

30. Ibid., p. 15.

31. Ibid., p. 8.

32. Goldring, W. 1927. The oldest known petrified forest. The Scientific Monthly, v. 24, No. 6, p. 515.

33. Dawson, J. W. 1871. On new tree ferns and other fossils from the Devonian. The Quarterly Journal of the Geological Society of London, v. 27. pp. 269–275.

34. Clarke, J.M. 1921. James Hall of Albany. Albany: Privately printed, p. 53.

35. Mather, W.W. 1843. Geology of New York, Part I, Comprising the First Geological District. Carroll and Cook, Albany, pp. 301–307.

36. Clarke, J.M. 1921. James Hall of Albany. Albany. Privately printed, p. 420.

37. Ibid., p. 421.

38. Sherwood, A. 1878. Section of Devonian rocks made in the Catskill Mountains at Palenville, Kaaterskill Creek, New York. Proceedings of the American Philosophical Society, v. 17, p. 347.

39. Prosser, C.S. 1899. Classification and distribution of the Hamilton and Chemung Series of central and eastern New York. Seventeenth Annual Report of the State Geologist for the Year 1897, Part 2, Wynkoop, Hallenbeck, Crawford Co., New York and Albany, p. 210.

40. Ibid.

41. Andrews, H.N. 1980. The Fossil Hunters. Cornell University Press, Ithaca, p. 229.

42. Prosser, C.S. 1899. Classification and distribution of the Hamilton and Chemung Series of central and eastern New York. Seventeenth Annual Report of the State Geologist for the Year 1897, Part 2, Wynkoop, Hallenbeck, Crawford Co., New York and Albany, p. 210.

43. Board of Water Supply of the City of New York. 1917. Catskill Water Supply — A General Description and Brief History, p. 7.

44. Ibid., p. 11.

45. Ibid., p. 9.

46. Ibid., p. 65.

47. New York Supreme Court — Schoharie Reservoir Petition. Schoharie County Archives, Schoharie, New York.

48. Ibid.

49. Ibid.

50. Board of Water Supply of the City of New York. 1917. Catskill Water Supply — A General Description and Brief History, p. 65.

51. Claimants' testimony in hearings involving the New York City Board of Water Supply and the residents of Gilboa. Schoharie County Archives, Schoharie, New York.

52. Ibid.

53. Ibid.

54. Ibid.

55. Ibid.

56. Schenectady Gazette, July 26, 1991.

57. Letter of Winifred Goldring to John M. Clarke, March 17, 1918. New York State Archives, Albany, NY.

58. Letter of John M. Clarke to Ruth Merrill, August 29, 1913. New York State Archives, Albany, NY.

59. Letter of Vincent Ayers to John M. Clarke, August 29, 1920. New York State Archives, Albany, NY.

60. New York State Museum — Paleontology Collection Locality Descriptions.

61. Goldring, W. 1924. The Upper Devonian forest of seed ferns in eastern New York. Eighteenth Report of the Director, New York State Museum Bulletin No. 251, The University of the State of New York, Albany, p. 51.

62. Ruedemann, R. 1926. Hunting fossil marine faunas in New York State. Natural History, v. 26, No. 5, pp. 509–511.

63. Goldring, W. 1924. The Upper Devonian forest of seed ferns in eastern New York. Eighteenth Report of the Director, New York State Museum Bulletin No. 251, The University of the State of New York, Albany, p. 51.

64. Ibid.

65. Ibid.

66. Letters of Winifred Goldring, 1927. New York State Archives, Albany, NY.

67. Letter of John M. Clarke to John H. Finley, President of the University of the State of New York, December 6, 1920. New York State Archives, Albany, NY.

68. Clarke, J.M. May 17, 1922. Report of the Department of Science and the State Museum. New York State Archives, Albany, NY.

69. Goldring, W. 1927. The oldest known petrified forest. The Scientific Monthly, v. 24, No. 6, p. 523.

70. Letter of Winifred Goldring to John M. Clarke, March 25, 1919. New York State Archives, Albany, NY.

71. Stewart, W.N. 1983. Paleobotany and the Evolution of Plants. Cambridge University Press, Cambridge, p. 245.

72. Letter of Edward W. Berry to Winifred Goldring. New York State Archives, Albany, NY.

73. Goldring, W. 1924. The Upper Devonian forest of seed ferns in eastern New York. Eighteenth Report of the Director, New York State Museum Bulletin No. 251, The University of the State of New York, Albany, p. 54.

74. Letter of Winifred Goldring to John M. Clarke, 1922. New York State Archives, Albany, NY.

75. Letter of Jules Henri Marchand to John M. Clarke, November 14, 1916. New York State Archives, Albany, NY.

76. Letter of Jules Henri Marchand to John M. Clarke, November 14, 1923. New York State Archives, Albany, NY.

77. Goldring, W. 1922. The oldest known petrified forest. The Scientific Monthly, v. 24, No. 6, p. 529.

78. Statement of acknowledgement of the restoration dedication by J. Austen Bancroft for Science, in a letter of Bancroft to John M. Clarke, February 23, 1925. New York State Archives, Albany, NY.

79. Letter of Redpath Museum, McGill University to Winfred Goldring, October 26, 1925 New York State Archives, Albany, NY.

80. Letter of A.C. Noe to Winifred Goldring, April 24, 1925. New York State Archives, Albany, NY.

81. Letter of F.A. Bather, Keeper of the British Museum (Natural History) to Winifred Goldring, November 29, 1926. New York State Archives, Albany, NY.

82. Letter of Riksmuseets, Paleobotaniska Avdelning, Stockholm, to Winfred Goldring, September 20, 1927. New York State Archives, Albany, NY.

83. Letter of Edward J. Foyles, Director of the Museum of Natural History, University of Rochester to Winifred Goldring, March 21, 1926. New York State Archives, Albany, NY.

84. Shipping memorandum of a stump sent to Professor Heinrich Ries, Geology Department, Cornell University. New York State Archives, Albany, NY.

85. Letter of the Geological Survey of Canada, Ottawa, to Winifred Goldring, March 26, 1926. New York State Archives, Albany, NY.

86. Memorandum of Professor Loomis of Amherst College to Winifred Goldring, December 21, 1928. New York State Archives, Albany, NY.

87. Letter of Thomas Johnson, Dublin, Ireland, to Winifred Goldring, February 3, 1927. New York State Archives, Albany, NY.

88. Letter of Margaret Ferguson of Wellesley College to Winifred Goldring, March 20, 1928. New York State Archives, Albany, NY.

89. Letter of Sidney K. Clapp to Winifred Goldring, June 30, 1927. New York State Archives, Albany, NY.

90. Ibid.

91. Letter of Sidney K. Clapp to Winifred Goldring, September 12, 1927. New York State Archives, Albany, NY.

92. Czerkas, S.M. and D.F. Glut. 1982. Dinosaurs, Mammoths and Cavemen — The Art of Charles R. Knight. E.P. Dutton, Inc., New York, p. 8.

93. Ibid., p. 29.

94. Letter of Charles R. Knight to Winifred Goldring, November 5, 1927. New York State Archives, Albany, NY.

95. Letter of Winifred Goldring to Charles R. Knight, November 10, 1927. New York State Archives, Albany, NY.

96. Andrews, H.N. 1980. The Fossil Hunters. Cornell University Press, Ithaca, p. 323.

97. Serlin, B.S., and H.P. Banks. 1978. Morphology and anatomy of *Aneurophyton*, a progymnosperm from the Late Devonian of New York. Palaeontographica Americana, 8: 343–359.

98 Ibid.

99. Ibid.

100. Ibid.

101. Banks, H.P. 1980. Floral assemblages in the Siluro–Devonian. *In* D.L. Dilcher and T.N. Taylor, eds., Biostratigraphy of Fossil Plants. Dowden, Hutchinson and Ross, Inc., Stroudsburg, p. 5.

102. Driese, S.G., C.I. Mora, and J.M. Elick. 1997. Morphology and taphonomy of root and stump casts of the earliest trees (Middle to Late Devonian), Pennsylvania and New York, USA. Palaios, 12:524–537.

103. Andrews, H.N. 1980. The Fossil Hunters. Cornell University Press, Ithaca, p. 205.

104. Arnold, C.A. 1937. Observations on fossil plants from the Devonian of eastern North America. III. *Gilboaphyton goldringiae*, Gen.et

sp. nov., from the Hamilton of eastern New York. Contributions from the Museum of Paleontology, University of Michigan v. 5, No. 7, pp. 75–78.

105. Grierson, J.D., and H.P. Banks. 1963. Lycopods of the Devonian of New York State. Palaeontographcia Americana, v. 4, No. 31.

106. Fairon-Demaret, M., and H.P. Banks. 1978. Leaves of *Archaeosigillaria vanuxemii*, a Devonian lycopod from New York. American Journal of Botany, 65(2): 246–249.

107. Berry, C.M., and D. Edwards. 1997. A new species of the lycopsid *Gilboaphyton* Arnold from the Devonian of Venezuela and New York State, with a revision of the closely related genus *Archaeosigillaria* Kidston. Review of Palaeobotany and Palynology, 96: 47–70.

108. Grierson, J.D., and H.P. Banks. 1963. Lycopods of the Devonian of New York State. Palaeotographica Americana, v. 4, No. 31.

109. Schuchman, P.G. 1969. *Pseudosporochnus* in the Middle Devonian of New York State. Unpublished M.Sc. thesis, Cornell University.

110. Skog, J.E., and H.P. Banks. 1973. *Ibyka amphikoma*, Gen. et sp. N., A new proto-articulate precursor from the late Middle Devonian of New York State. American Journal of Botany, 60 (4): 366–380.

111. Banks, H.P. Personal communication, April 6, 1995.

Chapter IV

1. Hendrix, L.E., ed. 1994. The Sloughters' History of Schoharie County. The Schoharie County Historical Society, p. 234.

2. Ibid.

3. Power Authority of the State of New York. 1973. Progress Toward Power.

4. Banks, H.P., P.M. Bonamo, and J.D. Grierson. 1972. *Leclercqia complexa* gen. et sp. nov., A new lycopod from the late Middle Devonian of eastern New York. Review of Palaeobotany and Palynology, 14:19–40.

5. Banks, H.P. Personal communication, April 6, 1995.

6. Bonamo, P.M. Personal communication, May 28, 1992.

7. Grierson, J.D., and P.M. Bonamo. 1979. *Leclercqia complexa*: Earliest ligulate lycopod (Middle Devonian). American Journal of Botany, 66 (4): 474–476.

8. Ibid.

9. Bonamo, P.M. 1977. *Rellimia thomsonii* (Progymnospermopsida) from the Middle Devonian of New York State. American Journal of Botany, 64 (10):1272–1285.

10. Dannenhoffer, J.M., and P.M. Bonamo. 1989. *Rellimia thomsonii* from the Givetian of New York: Secondary growth in three orders of branching. American Journal of Botany, 76 (9): 1312–1325.

11. Grierson, J.D., and H.P. Banks. 1983. A new genus of lycopods from the Devonian of New York State. Botanical Journal of the Linnean Society, 86: 81–101.

12. Bonamo, P.M., H.P. Banks, and J.D. Grierson. 1988. *Leclercqia*, *Haskinsia* and the role of leaves in delineation of Devonian lycopod genera. Botanical Gazette, 149 (2): 222–239.

13. Bonamo, P.M. Personal communication, May 28, 1992.

14. Ibid.

15. Shear, W.A. 1993. One small step for an arthropod. Natural History, March, pp. 47–51.

16. Selden, P.A., W.A. Shear, and P.M. Bonamo. 1991. A spider and other arachnids from the Devonian of New York and reinterpretations of Devonian Araneae. Palaeontology, 34 (2): 241–281.

17. Shear, W.A., P.A. Selden, W.D.I. Rolfe, P.M. Boanamo, and J.D. Grierson. 1987. A spider and other arachnids from the Devonian of Gilboa, New York (Arachnida, Trigonotarbida). American Museum Novitates, No. 2901, pp. 1–74.

18. Shear, W.A. 1993. One small step for an arthropod. Natural History, March, pp. 47–51.

19. Ibid.

20. Ibid.

21. Shear, W.A., and P.M. Bonamo. 1988. Devonobiomorpha, a new order of centiped (Chilopoda) from the Middle Devonian of Gilboa, New York State, USA, and the phylogeny of centiped orders. American Museum Novitates, No. 2927, pp. 1–30.

22. Ibid.

23. Ibid.

24. Shear, W.A., J.M. Palmer, J.A. Coddington, and P.M. Bonamo. 1989. A Devonian spinneret: Early evidence of spiders and silk use. Science, 246: 479–481.

25. Ibid.

26. Selden, P.A., W.A. Shear, and P.M. Bonamo. 1991. A spider and other arachnids from the Devonian of New York, and reinterpretations of Devonian Araneae. Palaeontology, 34 (2): 241–281.

27. Shear, W.A. 1993. One small step for an arthropod. Natural History, March, pp. 47–51.

28. Ibid.

29. Norton, R.A., P.M. Bonamo, J.D. Grierson, and W.A. Shear. 1988. Oribatid mite fossils from a terrestrial Devonian deposit near Gilboa, New York. Journal of Paleontology, 62 (2): 259–269.

30. Kethley, J.B., R.A. Norton, P.M. Bonamo, and W.A. Shear. 1989. A terrestrial alicorhagiid mite (Acari: Acariformes) from the Devonian of New York. Micropaleontology, 35 (4): 367–373.

31. Shear, W.A., and P.A. Selden. 2001. Rustling in the undergrowth: Animals in early terrestrial ecosystems. In P.G. Gensel and D. Edwards, eds., Plants Invade the Land. Columbia University Press, New York, p. 39.

32. Gray, J., and W.A. Shear. 1992. Early life on land. American Scientist, 80: 444–456.

33. Jeram, A.J., P.A. Selden, and D. Edwards. 1990. Land animals in the Silurian: Arachnids and myriapods from Shropshire, England. Science, 250: 658–661.

34. Dannenhoffer, J.M., and P.M. Bonamo. 1989. *Rellimia thomsonii* from the Givetian of New York: Secondary growth in three orders of branching. American Journal of Botany, 76 (9): 1312–1325.

35. Woodrow, D.L. 1985. Paleogeography, paleoclimate, and sedimentary processes of the Late Devonian Catskill Delta. In D.L. Woodrow and W.D. Sevon, eds., The Catskill Delta The Geological Society of America, Special Paper No. 201, pp. 51–63.

92-201PM

NOTES

NOTES